THE WORLDS OF GALILEO

The Inside Story of NASA's

Mission to Jupiter

Galileo
NASA
USA

THE WORLDS OF GALILEO

The Inside Story of NASA's Mission to Jupiter

MICHAEL HANLON

ST. MARTIN'S PRESS
NEW YORK

THE WORLDS OF GALILEO

For information, address St. Martin's Press, 175 Fifth Avenue, New York, N.Y. 10010

www.stmartins.com

ISBN 0-312-27220-0

First published in the United Kingdom by Constable, an imprint of Constable & Robinson Ltd.
First U.S. edition: November 2001

10 9 8 7 6 5 4 3 2 1

Contents

FOREWORD

"THE WORLDS OF GALILEO" was a title I used for Chapter 13 of *2010: Odyssey Two*, written almost twenty years before Michael Hanlon's account of one of the most successful space missions of all time. So it's not surprising that, as I followed the trials and tribulations of the real Galileo spacecraft, I found fact and fiction inexorably mixed – especially as one of the heroes of this long-drawn-out epic was Gentry Lee, my one-time collaborator.

It is hard not to draw comparisons with NASA's other cliffhanger – Apollo 13. True, that had the added ingredient of human drama, and was concentrated into a few days, whereas the Galileo mission extended over many years and it was only expensive hardware that appeared to be at risk. Yet perhaps even more important were the hundreds of man-years of devoted labour by some of the most brilliant engineers in the United States – which put a terrible strain on their health and personal relationships.

Although the four giant moons of Jupiter had been favourite telescopic objects ever since their discovery by Galileo in 1610, they were little more than points of light in the most powerful telescopes, and virtually nothing was known of their real nature when the space age dawned in 1957. However, it was not until the 1970s that rockets were powerful enough to send exploratory probes to Jupiter and beyond. The Pioneer, Voyager and Galileo spacecraft turned those dimensionless points into four fantastic worlds.

The series of setbacks, malfunctions and near-disasters that beset the Galileo mission would have seemed improbable in any work of fiction, and much of Michael Hanlon's book has the excitement of a good detective story. Incredibly, Galileo is still functioning, and returning megabytes of information from the worlds it has revealed to us in such detail.

But we still do not have the answer to the greatest question of all, though it is becoming more and more likely that one of these planet-sized moons may harbour life. Let me quote from my own "Worlds of Galileo":

> The unexpected discovery of life on Europa had added a new element to the situation – one that was now being argued at great length both on Earth and aboard *Leonov*. Some exobiologists cried "I told you so!" pointing out that it should not have been such a surprise after all. As far back as the 1970s, research submarines had found teeming colonies of strange marine creatures thriving in an environment thought to be equally hostile to life – the trenches on the bed of the Pacific. Volcanic springs, fertilizing and warming the abyss, had created oases of life in the deserts of the deep.
>
> Anything that had happened once on Earth should be expected millions of times elsewhere in the Universe; that was almost an article of faith among scientists. Water – or at least ice – occurred on all the moons of Jupiter. And there were continuously erupting volcanoes on Io – so it was reasonable to expect weaker activity on the world next door. Putting these two facts together made Europan life seem not only possible, but inevitable – as most of Nature's surprises are, when viewed with 20/20 hindsight.

Those were the words I wrote in 1980. I hope to know the truth well before 2010!

Arthur C. Clarke
Colombo, Sri Lanka
November 2000

Acknowledgments

The staff of the Jet Propulsion Laboratory in Pasadena have devoted an inestimable amount of their valuable time, energy and patience to telling me the story of Galileo – which in many cases is the story of their careers – and put me right a thousand times over. Mary Beth Murrill at the JPL Media Relations Office made the research for this book possible, and was always on hand to supply a vital phone number or email address. I thank John Casani, Neal Ausman and Norm Haynes for their invaluable insights into the mission and the early days, and also Bill O'Neil, Bob Mitchell, Phil Barnett, Matt Landano and Jim Erickson. I must give special thanks to Jim Marr, who spent hours explaining to me the complex repair mission carried out on Galileo's communications system. Jurrie van der Woude and Scott Chavez of the picture archive team brought the book to life. I must also thank Arthur C. Clarke for contributing the foreword – and for awakening my interest in Jupiter when in 1973 I first saw the movie *2001*.

I have spoken to dozens of scientists who have worked on the project; all have made a contribution to this book. Special mention must go, of course, to Torrence Johnson, and to Margaret Kivelson at UCLA. Thanks also to Rosaly Lopes-Gautier, who set me right on Io's volcanism; to Bob Pappalardo and Mike Carr, who steered me on a straight course through the oceans of Europa; and to Mike Belton, who explained to me the complexities of the Galileo Imaging System.

I must give special thanks to David Harland, whose encyclopedic knowledge of the history of space exploration has always been shared freely and with enthusiasm; to Carol O'Brien at Constable & Robinson, for her professionalism and commitment to this book; to John Woodruff for his diligence, skill and enthusiasm during the editing process; and to Anna Williamson for her patience and skill in putting the book together.

Last, and most of all, I thank Elena Seymenliyska, for her love, encouragement and enthusiasm, and for making all this possible.

Michael Hanlon
London
February 2001

Approximate metric-imperial conversions

All measurements in this book are given in metric units

Temperature

To change from Celsius to Fahrenheit, multiply by 9 and divide by 5, then add 32.

Length

One kilometre is about 0.62 miles.
One mile is about 1.6 kilometres.
One metre is about 3.3 feet.
One foot is about 0.3 metres.

Mass (weight)

One kilogram is about 2.2 pounds.
One pound is about 0.45 kilograms.

In this book "billion" means one thousand million (10^9), and "trillion" means a thousand billion (10^{12}).

The pictures in this book

False-colour images: Are we being fooled?

Many of the pictures in this book are shown in "false colour". This means that the Galileo imaging scientists decided to accentuate particular wavelengths by appropriate computer processing. This is *not* done to make the image prettier, or more dramatic. It can serve, for example, to highlight differences in neighbouring terrains, showing subtle landscape variations more clearly. Sometimes the colour change is slight – the picture on this book's cover shows Io pretty much as it would appear to the naked eye, with the yellows, greens, reds and oranges only slightly enhanced. Other pictures use false colour in a different way. For instance, the photograph of the Moon on page 48 uses colour to distinguish different rock types, and the resulting image, although beautiful, is clearly nothing like how the Moon appears in "real life". In this book (unlike in many other works) when a false-colour image has been used it is made clear in the caption.

Picture credits

All the photographs and artists impressions in this book are © NASA/JPL except the following:
All Hubble images are courtesy of the Space Telescope Science Intitute.
The pioneer graphic on page 9 is courtesy of Moreno Michini.
The graphics on pages 24, 60 and 61 © John Lawson.
The images of the Venusian surface in Chapter three are courtesy of the Russian Space Agency.

GALILÆUS GALILÆI.
Mathemat. Florentinus.

Chapter one

THE MEDICEAN STARS

And only this evening I have seen Jupiter accompanied by three fixed stars, totally invisible because of their smallness.

Galileo Galilei
Padua, 7 January 1610

IT IS HARD FOR us to imagine the wonder and astonishment filling the mind of Galileo Galilei as he spent the winter months of 1609–10 exploring the heavens with his new *instrumentum*, a spyglass consisting of two different-sized refracting lenses held together in a lead tube. Galileo did not, as is commonly believed, invent the telescope. Unusually for a thinker and academic heavyweight of his age, he always gave credit where it was due – in this case to the Dutch spectacle-maker Hans Lippershey, who had come up with the idea a couple of years before – but Galileo was certainly the first person to recognize the telescope's true potential.

Then, as now, securing funding for one's research took up a distressingly large proportion of the scientist's time. There were no funding councils in seventeenth-century Italy, but there was a great deal of money in certain quarters, and one of those was the Venetian senate. Galileo, who was very ambitious, heard about the Dutch telescope in the summer of 1609 from his friend Paolo Sarpi, a

This beautiful picture of Jupiter was taken by the Hubble Space Telescope, showing that above the confines of the atmosphere distant planets can be imaged with startling clarity.

Venetian theologian, and he immediately saw its potential. Galileo was not able to get hold of one of the new spyglasses, but, as perhaps the world's greatest expert on the science of optics, he grasped the principle quickly, and set about making his own. He also persuaded the Venetian rulers to hold off from awarding a contract to any rivals who might turn up with their own version. (The issue of a contract arose because this new piece of technology was perceived by those who held the reins of power to have military potential.) Galileo worked quickly, and by the end of 1609 he had the leaders of the Adriatic superpower eating out of the palm of his hand as they gazed in wonder from the towers and turrets of their city at the sails of ships far out at sea. His reward was a fat salary and a job for life at the University of Pisa.

Galileo went to work with his new device. The instrument he had presented to the Venetian senate magnified by a factor of eight. He now made a bigger, more powerful ×20 telescope and used it to scan the heavens. With this instrument he observed the Moon. Almost at once, everything changed. Months before, some other observers had pointed spyglasses at the lunar surface and commented on its bizarre appearance under high magnification, but Galileo was the first to see another *world* in the sky. He saw "protuberances and gaps" that "surpassed the Earth's", and correctly deduced that the Moon's surface was covered in mountains and valleys. He also saw "seas" which we now know to be dry basaltic floodplains – but this mistake was entirely forgivable: after all, no one knew then what an ocean would look like from space.

Galileo quickly realized that he had been given the keys to the cosmos. From his observatory in Padua, a city to the west of Venice, he could explore the Solar System and beyond. The Pleiades – the beautiful tight grouping of stars in the constellation Taurus, the Bull – was resolved into forty stars, where before there had only been six or seven visible to the naked eye. The great Hunter, Orion, suddenly became some five hundred stars richer, and Galileo was the first to see the Milky Way for what it is: "nothing but a congress of innumerable stars grouped together in clusters".

But his greatest discovery took place on 7 January 1610, a discovery that destroyed for ever the old idea of the Solar System as a series of fixed spheres. Pointing his telescope at Jupiter, the largest and brightest planet in the sky apart from Venus, he saw that the brilliant, yellow-white blob in the sky resolved itself into a disc. Suddenly it was looking as though Jupiter might also be a world, just like Earth and its Moon. But it was what he saw just to the side of Jupiter that really surprised him – three tiny "stars". The following evening he pointed his telescope at Jupiter again. The first evening, two of the "stars" were east of Jupiter and one to the west. By the following day the arrangement was again different – now all three were to the west. Initially he thought that Jupiter itself was zigzagging across the sky. Further observations showed the "stars" in different positions still, and furthermore their brightness seemed to vary from day to day. It suddenly occurred to him that it might not be Jupiter that was moving in such a strange way (after all, this would surely have been noticed before, even in the pre-telescopic age). The movement of these "stars" could be better explained by supposing that they were moving around Jupiter.

On the twelfth day he saw only two stars, but on the 13th he saw a fourth. "Three were to the west and one to the east", he wrote. He concluded that he was looking at a group of Jovian satellites. Not only was Jupiter a world in its own right, but this world had its own family of four moons – and, like our Moon, they too were presumably worlds in their own right.

Galileo rushed to publish an account of his findings – after all, telescopes were becoming commonplace and it could only be a matter of time before somebody else had the idea of training a spyglass on mighty Jupiter. On 13 March 1610 Galileo's *Sidereus nuncius* ("The Starry Messenger") was published, in Latin. It caused a sensation. Galileo called the Jovian satellites the Medicean Stars in honour of Cosimo de' Medici, a member of the Tuscan noble clan that wielded much power and influence. Soon, they were being referred to as the "Galilean satellites" in honour of their discoverer, but in 1614 they were given individual names: in order of increasing distance from Jupiter, Io, Europa, Ganymede and Callisto – the worlds of Galileo.

(The names were the choice of Simon Marius, a German astronomer who independently discovered the moons of Jupiter at about the same time as Galileo.)

In this 'family portrait' montage, the four Galilean Satellites are shown to scale. From left to right are Io, Europa, Ganymede and Callisto. These global views show the side of volcanically active Io which always faces away from Jupiter, icy Europa, the Jupiter-facing side of Ganymede, and heavily cratered Callisto.

A new world order

To appreciate the full impact of Galileo's discoveries it is necessary to imagine the mindset of the vast majority of people alive at that time. Although for centuries the educated classes had known basic things about the Universe such as the fact that the Earth was a sphere, among the masses the idea that their world was a large ball of rock and water floating in an endless void was at best a hazy and bizarre concept. To suggest that the points of light in the sky were also globes in their own right would be so strange as to have been beyond most people's imagining. The first theory of gravity was still decades away; people still struggled with the concept of objects adhering to the lands of the southern hemisphere, let alone the far side of the Moon. The Church, which through its priests was still the only contact most common people had with the educated classes, had always firmly resisted any cosmology that conflicted with the fiercely geocentric dogma of Claudius Ptolemy. Putting the Earth at the centre of things made the Universe, and its God, deeply personal and relevant to the human race. The trouble was, the evidence was stacking up against this cosy world-view. In the 1540s, seventy years before Galileo saw the light at Jupiter, the Polish genius Nicholas Copernicus announced that the Sun, not the Earth, was the centre of the Universe (or of the Solar System at least). Copernicus made a good stab at proving, mathematically, that the motions of the planets, the Sun and the Moon could be best explained by assuming that the planets orbited the Sun and the Moon orbited the Earth. He had no idea how this occurred, *why* these huge spheres should wish to travel in endless circular paths (that had to wait for the attentions of Mr Newton), but circle they did.

From our modern perspective we can safely dismiss the old geocentric ideas of Ptolemy, and we may even deride the superstitious Christian Church for clinging to the teachings of Ptolemy and Aristotle as the basis of its world-view, in the face of evidence to the contrary. But in Ptolemy's day this model of the Universe made perfect sense – after all, why *not* have everything revolving around the Earth? In the end it was maths, not faith, that was the undoing of the old Earth-centred brigade. Copernicus showed that the strange meandering motion of the planets (the word *planet* means "wanderer" in Greek) could be best explained if they were all – including the Earth – revolving around the Sun. The problem with this view was that the Moon most definitely revolved around the Earth. Why should the Earth alone have the privilege of possessing a satellite? Galileo's discovery that Jupiter – another planet – is not moonless, and has not one but four moons, immediately relegated the Earth to just another sphere.

Galileo's account of his discoveries – an incredibly short one at just 24 pages – sold out immediately. *Sidereus nuncius* had a similar impact on Europe's intelligentsia that Charles Darwin's *Origin of Species* was to have more than two centuries later. Later, the British academic William Lower wrote, "Me thinkes the diligent Galileus hath done more in three fold discourie than Magellane in openinge the streightes to the South Sea." Eventually Galileo paid the price for all this – his run-in with Rome almost earned him a painful session with the Inquisition, and in the end he suffered the indignity of house arrest.

In the course of a few evenings staring up at the chilly, clear north Italian skies with his primitive telescope, less powerful than a decent pair of binoculars today, Galileo caught a glimpse of the true nature of the Cosmos. There were strange worlds out there: globes covered with mountains, valleys and, apparently, oceans. Of the four worlds he discovered orbiting Jupiter, Galileo knew nothing, save their separation from the great planet and how long they took to complete one revolution around their master. His telescopes were not powerful enough to reveal the Jovian moons as any more than points of light. But in the coming centuries, bigger and better spyglasses would be trained on these worlds and, in time, humans would send robotic emissaries to them across the void. Galileo would have been amazed to learn just how spectacular Io, Europa, Ganymede and Callisto would turn out to be.

Closing in on the worlds of Galileo

Although they remained mere points of light until the late 1840s, the Galilean satellites became important tools for the astronomer. Galileo himself was the first to see one great potential of his discovery – solving the longitude problem. The seventeenth century was a golden age of European exploration. Scarcely a year went by without a fleet of ships setting sail from England, Spain, Holland or Portugal, their crews hoping to find new lands whose treasures and bounties were just waiting to be grabbed. The problem was that navigating in those days was more an art than a science. Position north or south of the equator – one's latitude – could be determined accurately using a sextant. But working out how far east or west you were of a certain point was quite another matter. For this you need some way of measuring the passage of time, accurately and consistently – to an accuracy of seconds per day. And in the early 1600s accurate clocks that could be carried aboard ships and keep good time on stormy seas simply did not exist.

Galileo had a bright idea. He knew that eclipses of the four Galilean satellites occurred around a thousand times a year, and so predictably that you could set a clock by them. A highly accurate indicator of time already existed – up in the sky. To solve the longitude problem all you needed was a telescope. He presented his plan to King Philip II of Spain, who was offering a small fortune – a substantial lifetime stipend – to the "discoverer of longitude". (These were the days when the great and the good took an interest in the latest developments, and saw science as an investment, not a burden of subsidy.) But the king had been worn down by a thousand cranky ideas from would-be claimants of the prize. His committee of experts set up to award the prize rejected Galileo's plan, making some perfectly valid objections. First, the "Jupiter clock" would only work at night, and in clear weather. Second, making accurate observations of the planet and its tiny moons from the surface of a pitching and tossing ship would tax the steady hands of the best astronomer, let alone a sailor. Galileo himself conceded that the pounding of a sailor's heart would be enough to send Jupiter streaking out of the field of view of a hand-held telescope, and that was without taking the effect of waves into account. And third, even in clear weather and at night, Jupiter is not always visible.

Galileo touted his idea around the various seafaring powers. The Tuscan Government was offering a similar prize, as was the Dutch. Both rejected his plan for the same logical reasons as the Spanish, but at least the Dutch gave him a thick gold chain for his troubles. Later, after Galileo's death, the British would offer the greatest longitude prize of all, to be claimed by the Yorkshire clock-maker John Harrison in 1773.

Until Harrison set to work, explorers had to rely on the skies to tell them where they were. Galileo's method finally became accepted in the decade after his death in 1642. Surveyors found that the Jupiter clock worked well, but only on land, and then only in the hands of skilled surveyors who were trained in the art of using a telescope and making accurate astronomical observations. The maps of the continents were soon redrawn. Galileo's four little moons were responsible for knocking the world more or less into the shape familiar to us today. In 1668, Giovanni Domenico Cassini published the definitive tables of the Jovian eclipses, which led to the further redrawing of maps and the redefining of empires. Distant Io, Europa, Ganymede and Callisto were proving to have a great influence here on Earth.

Jupiter's moons had yet more uses. In 1675 the Danish astronomer Ole Römer used eclipse timings for the Galilean satellites to show conclusively that light did not travel instantaneously, as had been assumed. As Jupiter and the Earth move round the Sun in their respective orbits, the distance between the two planets varies greatly, from around 630 million to nearly a billion kilometres. The eclipses of all four of the Jovian satellites would occur ahead of schedule when the Earth came closest to Jupiter, and behind schedule when the two planets became further apart. Römer used the discrepancies to make the first measurement of the speed of light – a slight underestimate, as it turned out, but still close to the accepted modern value of about 300,000 km per second. On many occasions Galileo himself had tried – and failed – to measure light speed (he suspected quite rightly that nothing travels infinitely fast), but despite years of clambering up and down Italian mountains with clocks, mirrors and lanterns he never managed to get an answer. It took the immense interplanetary yardstick stretching between the Earth and Jupiter's moons to give a true reading of the Universe's ultimate speed limit.

By the mid-nineteenth century, telescopes were finally getting good enough to resolve Galileo's "stars" into mini-planets in their own right. Two of them, Callisto and Ganymede, turned out to be planet-sized – bigger than Mercury. Io and Europa were similar in size to Earth's Moon. The first feature to be detected on any of the Galileans was a "polar spot" seen on Ganymede in the late nineteenth century: US astronomers Edward Emerson Barnard and Andrew Ellicott Douglass both claimed they had seen markings on Ganymede in about 1898, though the "canals" described by Douglass were disputed by Barnard, who said he saw more vague and generalized shading. In 1908 the *Astronomische Nachrichten* published a sketch of Ganymede which the Spanish astronomer José Comas Solá had made the previous year. By the 1960s, reasonably accurate maps of Ganymede were in existence. At the dawn of the space age, the sizes and masses of Jupiter and its four main moons had been ascertained. In addition, spectroscopic analysis of Jupiter had revealed its composition – mostly hydrogen and helium, with a smattering of methane and ammonia. Astronomers had discovered that Jupiter is covered with a swirling mass of clouds, and some of the largest weather patterns, notably the Great Red Spot – an apparently permanent feature that could swallow the Earth with plenty of room to spare – had been charted. But to learn any more about Jupiter and its empire, humanity was going to have to get up close.

The Pioneers

The first spacecraft to be sent to the outer Solar System were two tiny robotic probes built by NASA's Ames Research Center in Mountain View, California. Launched on 2 March 1972, Pioneer 10 was the first

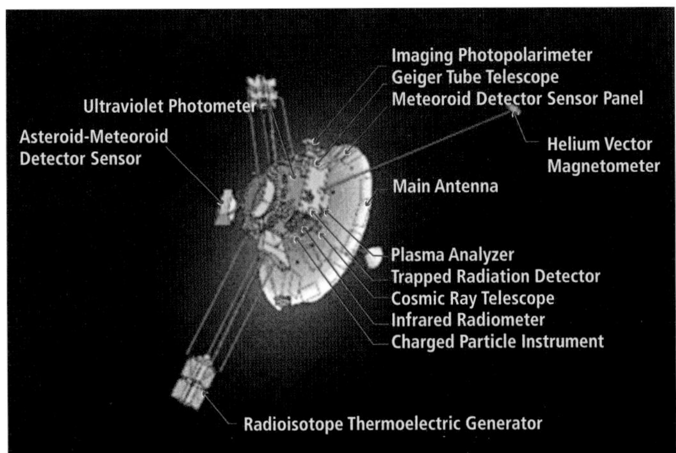

LEFT The two Pioneers of the early 1970s broke new ground in space exploration. Visiting Jupiter and then Saturn, they opened up the Outer Solar System to later exploration – detecting the strong radiation belts around Jupiter that have the power to cripple a spacecraft's electronics unless adequately protected.

BELOW RIGHT The Pioneer spaceprobes were equipped with only rudimentary imaging systems, and did not have a stable scan platform from which to take the crystal-clear images captured by the later Voyagers and Galileo. Nevertheless, these two spacecraft, humankind's first emissaries to the Outer Solar System, did manage to send back some impressive colour pictures. This is Jupiter, taken by Pioneer 10 in 1973. The black disc is Io, which is between the spacecraft and the planet.

spacecraft to travel through the asteroid belt and the first to make direct observations and obtain close-up images of Jupiter. Pioneer 10 has achieved fame as the most remote object ever created by humans – it is currently about 11 billion km away. The spacecraft made valuable scientific investigations in the outer regions of our Solar System until the end of its mission on 31 March 1997. Pioneer 10 is headed in the direction of the constellation Taurus. It will take Pioneer over 2 million years to encounter its first star.

Launched 13 months later, Pioneer 11 followed its sister ship to Jupiter (1974) and made the first direct observations of Saturn in 1979. The Pioneer 11 mission officially ended on 30 September 1995, when the last transmission from the spacecraft was received. Pioneer 11 is travelling in the direction of the constellation Aquila, the Eagle – it may rendezvous with one of the stars there in about AD 4,000,000.

Compared with what came later, the Pioneers were fairly primitive, but nevertheless achieved an incredible amount of science. Both were spinning "particles and fields" probes, designed to explore the radiation and magnetospheres (those regions of space dominated by a planet's magnetic field) of the outer planets, and the solar wind that dominates interplanetary space. One of their main achievements at Jupiter was to detect the intense radiation storm that surrounds the planet. This was vital information because knowing how hostile the Jovian environment is, NASA's engineers were able to take measures to protect the delicate electronics of later space probes.

Although the Pioneers did not have true cameras, they were able to take pictures of Jupiter as they rushed past, using their photometers. Although crude and of poor resolution compared with what was to come, these were the first close-up pictures of the planet, and rightly earned the spacecraft their place on the front pages of the world's newspapers. The Pioneer views of the Galilean satellites revealed little in the way of surface detail, but did show that Ganymede's surface was composed of distinct light and dark areas, that Europa shone like a mirror, and that Callisto was dark and cratered.

JPL – a new era in space exploration

Scattered across the foothills of the San Gabriel mountains a few kilometres north of the pleasant Los Angeles suburb of Pasadena, lies the Jet Propulsion Laboratory (JPL), NASA's planetary exploration and space science centre. JPL has a long and rather muddled history, with quasi-military beginnings. It was originally set up to make rockets for the military. Then it became subsumed into NASA, although its employees are on the payroll of the California Institute of Technology (Caltech). Driving into the vast complex, you get a sense of travelling into the past. Road signs bearing the futuristic NASA logo look straight out of the 1960s. Many of JPL's buildings are utilitarian, and the whole place is painted in a rather off-putting beige. JPL is a strange amalgam of casual, laid-back science campus and top-secret military research centre – foreign visitors still have to have their passports scrutinized at the gates and must be accompanied by a JPL employee at all times.

Early planetary exploration was the responsibility of JPL, and the labs had designed, built and flown the successful Mariner spacecraft to Venus and Mars, which sent back spectacular pictures in the 1960s and

The Jet Propulsion Labs nestle in some of the most scenic real estate in the Greater Los Angeles area. Behind are the imposing San Gabriel Mountains.

early 1970s. JPL then came up with an ambitious plan for a "grand tour" of the outer Solar System, using an extensively updated variant of the Mariner design to explore Jupiter, Saturn, Uranus and Neptune. In the end, the two craft that resulted from this plan – Voyager 1 and Voyager 2 – were rather scaled down in ambition but still represented a huge advance on the Pioneers. For a start, both had state-of-the-art cameras on board, as well as sophisticated particle and field experiments, and infrared spectrometers to measure planetary composition, all mounted on an articulated "scan platform". Radioactive batteries powered the Voyagers – a system that has been used on all outer Solar System missions. Ideally, NASA would like to use politically correct solar panels, but beyond the orbit of Mars there just isn't enough sunlight to power a robotic space probe.

In 1972, when the Congressional committee that oversees NASA spending gave the Voyagers the thumbs-up, it was considered an act of foolhardiness not to back up every system on a spacecraft just in case something went wrong. The ultimate "redundancy strategy" was to back up the entire probe. So two Voyagers were built and were launched in 1977 using Titan IIIE-Centaur rockets, then the most powerful launch system at America's disposal. They weighed nearly a tonne

The Voyager spacecraft. Many of the systems and components used on the Galileo spacecraft were derived from those on the two Voyager probes. Like Galileo, the Voyagers had sophisticated on-board computers, "hardened" against the harsh radiation in the Jovian environment. Like Galileo, the Voyagers were powered by on-board radioactive generators, rather than solar panels, which would not work well in the feeble sunlight beyond the orbit of Mars. Artist's impression (LEFT) and diagram showing component parts (BELOW).

apiece, and were by far the most ambitious and sophisticated robotic spacecraft ever flown. But they were good value: at a total cost of $865 million in 1970's money, including design, build, operation and launch costs, the Voyager double mission came to about 20 cents per US citizen per year of operation – literally "peanuts".

Both Voyagers were specifically designed and protected to withstand the large radiation dosage they would receive during the Jupiter flyby. This was accomplished by using radiation shielding and selecting components designed to withstand a high level of ionizing radiation. An unprotected human passenger riding aboard a Voyager during the Jupiter encounter would have received enough radiation to kill him a thousand times over.

At Jupiter, the two Voyagers revealed a mini solar system of exquisite beauty and strangeness, alien vistas that rank among the most stunning twentieth-century views of any kind. Neither spacecraft was able to dwell at Jupiter. Voyager 1 swept by on 5 March 1979; Voyager 2 arrived on 9 July. Both spacecraft had started taking pictures of the planet months before their close encounter, from distances of millions of kilometres, and even these images were better than anything taken from Earth. Voyager 1 completed its Jupiter encounter in early

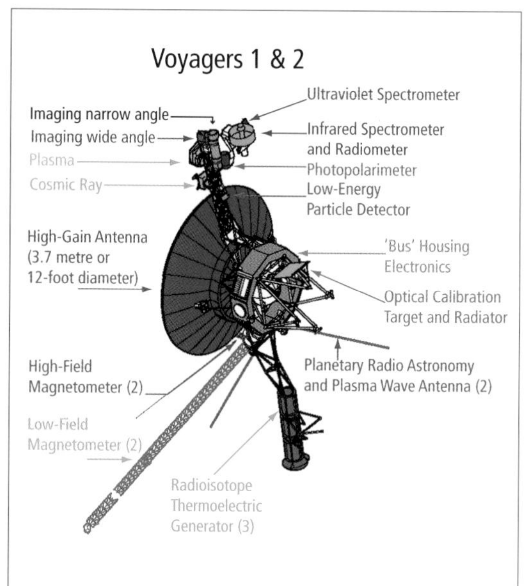

April, after taking almost 19,000 images and making many other scientific measurements. Voyager 2 picked up the baton in late April and its encounter continued into August. Together, the two probes returned more than 33,000 images of Jupiter and its satellites.

The Voyager discoveries

Although astronomers had studied Jupiter from Earth for several centuries, scientists were astounded by the Voyager findings. When it came to understanding the dynamics of Jupiter's atmosphere and the processes operating on its moons, it was back to the drawing board. Discovery of active volcanism on the satellite Io was probably the greatest surprise. It was the first time active volcanoes had been seen on another planet. Io appears to be the primary source of the matter that pervades the Jovian magnetosphere. Sulphur, oxygen and sodium, erupted by Io's volcanoes and sputtered off the surface by the impact of high-energy particles, were detected at the outer edge of Jupiter's magnetosphere.

The beautiful and surreal Voyager images of the delicate swirls and eddies of Jupiter's atmosphere were a combination of art and science. Nothing like this had ever been seen before – titanic clouds, swirling storms, eddies and hurricanes the size of the Earth's continents. Jupiter has weather, and a lot of it. The question is, is it like the weather we are familiar with on Earth, driven by the Sun, or is Jupiter's meteorology driven by a completely different engine? The Voyagers saw cloud-top lightning bolts similar to those that flash high in the Earth's atmosphere. They measured the temperature at Jupiter's cloud-tops – about -130 to -110°C. The spacecraft also measured the bulk composition of the upper atmosphere, finding a ratio of hydrogen to helium of around 88:11.

The Voyager mission's biggest surprise came at Io. Voyager 1 identified nine erupting volcanoes on Jupiter's innermost moon. Voyager 2 observed eight of the nine; the largest had shut down by the time it arrived at Jupiter. Plumes from the volcanoes reach more than 300 km above the surface. The material was being ejected at velocities of a kilometre a second. By comparison, lava at Mount Etna, one of Earth's most explosive volcanoes, is only thrown out at a mere 50 m per second.

Europa was a puzzle. The Voyager pictures showed a large number of mysterious linear features. These were so lacking in topographic relief that they "might have been painted on with a felt marker", as one scientist commented at the time. These bizarre lines and cracks on Europa's surface defied explanation. Clearly, NASA was going to have to go back and take another look.

Ganymede was seen to have two distinct terrain types – cratered and grooved, telling scientists that Ganymede's entire, ice-rich crust has been under tension from global tectonic processes. Callisto, the Voyagers found, has an ancient, heavily cratered crust, with shadowy rings – the ghost-like remains of enormous impact basins. The largest craters were apparently erased when the ice-laden crust flowed over periods of hundreds of millions of years.

Voyager also discovered a ring around Jupiter. Its outer edge is 129,000 km from the centre of the planet, and, though the brightest portion is only about 6,000 km wide, ring material may extend another 50,000 km down to the top of Jupiter's atmosphere. The ring is no more than 30 km thick. We now know that all the gas giants – Jupiter, Saturn, Uranus, and Neptune – possess rings – although each ring system is unique and distinct from the others.

Voyager detected an electric current of 5 million amps arcing between Jupiter and Io. As Io sweeps through Jupiter's magnetic field, it generates an electric potential linking the two bodies. Voyager discovered that Jupiter howls and squawks at radio frequencies. Radio blasts between 10 kilohertz and 1 megahertz may result from plasma oscillations in the Io torus – the doughnut-shaped ring swept by Io as it orbits Jupiter.

Voyager – still up and running

A total of five trillion bits of scientific data had been returned to Earth by the two Voyagers by the time they passed the orbit of Neptune. This represents over 6,000 complete sets of the *Encyclopaedia Britannica*. The Voyager spacecraft will be the third and fourth human artefacts to escape entirely from the Solar System. Pioneers 10 and 11 both carried small metal plaques identifying their time and place of origin for the benefit of any alien spacefarers that might find them in the distant future.

NASA placed a more ambitious message aboard Voyagers 1 and 2 – a kind of time capsule intended to communicate the story of our world to extraterrestrials. The Voyager message is carried on an analogue phonograph record – a 30 cm gold-plated copper disc containing sounds and images selected to portray the diversity of life and culture on Earth. The contents of the record were selected by a committee chaired by the late Carl Sagan of Cornell University, the astronomer and popularizer of science. Sagan and his associates assembled 115 images and a variety of natural sounds, such as those made by surf, wind and thunder, birds, whales, and other animals. To this they added musical selections from different cultures and eras, spoken greetings in fifty-five languages, and printed messages from President Carter and United Nations Secretary General Kurt Waldheim. Each record is encased in a protective aluminium jacket, together with a cartridge and a needle with which to play it. Instructions, in symbolic language, explain the origin of the spacecraft and indicate how the record is to be played.

The Voyagers left the realm of the planets in the early 1990s. Now they are in empty space, although still very much in contact with their masters back on Earth. It will be 40,000 years before they make a close approach to any other star. As Carl Sagan said at the time, "The spacecraft will be encountered and the record played only if there are advanced spacefaring civilizations in interstellar space. But the launching of this bottle into the cosmic ocean says something very hopeful about life on this planet."

Return to Jupiter

As far back as the 1960s, scientists at NASA were thinking about a dedicated Jupiter space probe. Jupiter was clearly an interesting place. Even then, back-of-the-envelope calculations showed that it should be possible to get a spacecraft weighing a tonne or so out to Jupiter, carrying enough propellant to slow it down when it got there and drop into orbit.

In 1975, the Ames Research Center started developing a Pioneer Jupiter Orbiter Probe. It would be a two-part spacecraft – an orbiter that would study Jupiter and its moons from space, and an atmospheric probe which would be released from the main craft and plunge into Jupiter's atmosphere. At the same time JPL was planning a similar mission. In 1982 NASA decided that JPL should undertake all future planetary missions, and the Jupiter mission transferred to Pasadena.

The Jupiter Orbiter-with-Probe (JOP) mission was sketched out in the mid-1970s, before the Voyagers had even arrived at Jupiter. The mission design went through many incarnations, and had to take into account NASA's new launch vehicle – the Space Shuttle. On the one hand, the Space Shuttle was a good thing because its roomy cargo bay and prodigious lifting ability meant that a truly gargantuan probe could be carried aloft. On the other hand, it was a very bad thing because although JPL was committed to using the Shuttle for its Jupiter mission, the new spaceplane had not actually flown yet.

Project Galileo, as the JOP was christened, was given formal Congressional approval on 1 October 1977, nearly two years *before* the Voyagers arrived at Jupiter. Like any multi-billion-dollar space mission, Galileo had many progenitors, but if there is anyone who can lay claim to be the father of the mission it is John Casani, now formally retired but still very much a larger-than-life presence in the corridors and offices of JPL. Casani, together with Gentry Lee (who has become a successful science-fiction writer) and other key figures, such as James Van Allen, were responsible for a mission that would dominate several hundred people's lives for a quarter of a century.

Casani takes up the story: "I was in charge of Galileo for 10 years, 4 months and the better part of a week if I remember correctly. At that time we were in the middle of the development of the Shuttle and the spacecraft mission was proposed to be one of the earlier flights. As far as the Galileo mission, or JOP as it was called at the time, the Congressional debate peaked up in the summer of 1977. However, the House of Representatives gave a 'no' vote on it, then they went to the Senate, then the Senate Committee approved it. That meant that the issue had to be resolved in conference between the members of the House of Representatives and the members of the Senate. In conference they

The pioneer images of Jupiter and Saturn caught the public's imagination. But far, far better was to come from the two Voyager probes. Here is Jupiter's Great Red Spot, surrounded by the swirling clouds and eddies of the upper atmosphere. For the first time the cloudbelts seen from Earth had been resolved into a fantastic, kaleidoscopic panorama of alien beauty. Surrounding the Red Spot – which was first seen in the 18th Century – are numerous lesser "white spots", temporary storms the size of Asia.

agreed to take the issue back to the House for approval after some discussion, and at that time there was a very positive and very supportive vote. In the end the vote was pretty lopsided, 250 to 100 or something like that. So we got in the end a very politically strong endorsement, after very shaky initial deliberations."

So what was to become the Galileo project was underway. All being well, the new Shuttle would be ready to blast the spacecraft off to Jupiter in 1982. Galileo would arrive at Jupiter in 1985, after getting a look at Mars on the way. It would enter orbit around the giant planet, taking pictures and sending back gigabytes of data on Jupiter and its moons, as well as data relayed from the atmospheric probe. But all was not to be well. Like the political deliberations, the future of Galileo was to prove very shaky indeed.

On 8 February 1992 another interplanetary spacecraft, Ulysses, swept past Jupiter's orbit after a voyage of just 16 months. Ulysses – a joint NASA–European Space Agency probe – was not designed primarily to study Jupiter. Its aim was to explore interplanetary space at high solar latitudes – far out of the plane in which the planets orbit. Nevertheless, its particle and field instruments sent back valuable data on Jupiter's magnetic field.

Jupiter and its moons – what we knew before the Galileo mission

Through a telescope Jupiter appears as a yellowish disc crossed with orange-red bands. Since the mid-1600s, astronomers have noted "spots" moving across Jupiter's face as the planet rotates. Some of these spots have survived for decades; one, the Great Red Spot, has persisted for nearly two centuries. Jupiter is a world of superlatives. Jupiter, along with Saturn, Uranus and Neptune, is a so-called gas giant. Unlike the rocky terrestrial planets – Mercury, Venus, Earth and Mars – the gas giants have no hard surface, although all have small solid cores. Pluto, a ball of ice and frozen gas, is in a category all of its own.

Jupiter is huge. At 142,984 km across at the equator, it is bigger, and more massive, than all the other planets, asteroids comets and moons in the Solar System combined. Jupiter could swallow 1,325 Earths, yet because it is made mostly of hydrogen and helium rather than rock, it weighs "only" 318 times as much as our planet. Sending a probe into Jupiter's atmosphere is a one-way mission. Jupiter's gravity is 2.69 times the Earth's, giving an escape velocity of some 60 km per second. Jovian atmosphere sample-and-return missions are likely to be off the agenda for a long time yet. Despite its size, Jupiter spins like a top. Its day is under 10 hours long, meaning that the rotational speed of a point on the equator is about 45,000 kph! This forces Jupiter to bulge out at the equator – its polar diameter is 8,000 km less than that at the equator. Jupiter's bulk composition is 81 per cent hydrogen and 18 per cent helium. The remaining 1 per cent – nearly four "Earths-worth" – is a mixture of methane, ammonia compounds, and heavier elements like carbon, sulphur, oxygen and silicon.

Jupiter's weather displays amazing similarities to that of Earth, despite its size and alien composition. The planet is swept by about a dozen prevailing winds, reaching 150 metres per second at the equator. On Earth, winds are driven by the large differences in temperature – more than 40°C – between the poles and the equator. This temperature difference is accounted for entirely by differences in solar heating between high and low latitudes. But Jupiter's clouds are roughly the same temperature – about -130°C – at both pole and equator. This is about 40 degrees colder than the lowest temperatures ever recorded in Antarctica – but not as cold as would be expected given Jupiter's distance from the Sun – about 778 million km, or just over five times the distance from the Earth to the Sun. Jupiter evidently produces heat in its own right – and a lot of it. One of Galileo's key tasks was to find out if the weather seen at Jupiter's surface is driven by the Sun, as on Earth, or by the furnaces below.

Moving on to the satellites, Jupiter had, it was thought at the time of Galileo's launch, 16 moons, only four of which – the Galilean satellites – are worlds of any substance. The Galileans – Io, Europa, Ganymede and Callisto – range in size from slightly smaller than our Moon to slightly larger than Mercury. The largest is Ganymede; at 5,268 km across it is the biggest satellite in the Solar System and a world substantial enough to be a planet in its own right. Ganymede's surface is made largely of water ice (as we know from spectroscopic measurements made from Earth). Ganymede orbits Jupiter at a distance of just over a million kilometres. The innermost moon, Io, is smaller, at 3,634 km across, and orbits Jupiter at a distance of just 421,600 km. Io's surface is ice-free, and this moon is dense – at 3.53 grams per cubic centimetre. This means that Io must be made largely of rock and metal. Interestingly, Io is also the reddest object in the Solar System, redder even than Mars, making it the true Red Planet. Next comes Europa. Brilliant white, little Europa, the smallest of the Galileans at just 3,136 km across, looks almost featureless from Earth. Europa orbits Jupiter at 670,000 km, once every 3.55 Earth days. Europa's surface appears, from Earth, to be composed almost entirely of pure ice, although its density of 2.97 grams per cubic centimetre suggests that only its crust can be made of frozen water; underneath must lie a world composed of silicates and metals. Ganymede is the third moon, and then comes Callisto. At 4,806 km across, Callisto is the second largest of the Jovian satellites. Voyager photographs showed a cratered surface, superficially very similar in appearance to our own Moon. Callisto's density, at just 1.86 grams per cubic centimetre, implies that it is composed largely of ices, probably mixed with rocky fragments. Callisto is the least dense "rocky" planet or moon in the Solar System.

As well as the four large Galilean satellites, easily visible from Earth, Jupiter has an additional 12 moonlets, four inside the orbit of Io, and eight outside the orbit of Callisto. They range in size from Amalthea, which is a peanut-shaped rock 131 km long, to tiny Leda, which is less than 5 km – the smallest satellite of any planet (unless you count the chunks of ice and rock that make up Saturn's rings). S/1999 J1 will be given a "proper name" in due course.

The new Jupiter probe had a hard job ahead of it. It had to get to Jupiter, slow down and enter orbit. It then had to spend many years photographing, probing, tasting and divining the worlds discovered by Galileo all those centuries before. If anyone could achieve all this, it was the scientists and engineers at JPL.

Chapter two

JPL GETS TO WORK

FLYING MORE THAN A billion kilometres across the Solar System, stopping when you get there, then parking yourself in an orbit around a huge planet 143,000 km across and undertaking a complex survey of that planet and its moons (as well as sending the results back to Earth) is no easy undertaking. That people were even talking about a Jupiter probe as early as the 1960s – when the space age was less than a decade old – almost beggars belief. It is like proposing to stage a Formula One race a few years after the invention of the wheel. But in those heady days, when NASA was working on a tight ten-year schedule to get a man on the Moon, anything seemed possible.

The design brief was challenging, to say the least. Galileo had to get to Jupiter in a reasonable amount of time. Zooming to Jupiter on a direct trajectory – "as the crow flies" – would require a lot of fuel. To save fuel, the spacecraft could be directed to travel via one or more planets, such as Mars, using their gravities to accelerate Galileo like a slingshot as it swung past. So some sort of tour around the Solar System would be needed, calling upon the gravitational assistance of at least one of the other planets. The probe would have to carry its own power supply and have enough on-board computing power to be autonomous: although it would be under the command of mission controllers on Earth, radio signals take over an hour to get to Jupiter and back, so making last-minute course adjustments to avoid impending disaster would not be possible.

The probe had to survive the harsh environment around Jupiter. There's the cold to contend with, and the vacuum of deep space – which can spot-weld exposed metal surfaces together in seconds – but also the space around Jupiter is also fiercely radioactive. Thanks to the planet's enormous magnetic field, and its interaction with the solar wind, Jupiter sits in a lethal radiation storm capable of frying an astronaut in hours and, more importantly for Galileo, a computer in minutes. Stray subatomic particles – bits of atom travelling at nearly the speed of light – are quite capable of reprogramming a silicon chip to do nothing whatsoever. Any electronics aboard the Jupiter spacecraft would have to be "hardened" against the fierce Jovian radiation.

Then there were the requirements of the mission itself. The spacecraft would have to scrutinize Jupiter and its moons. There would have to be cameras to take close-up pictures of the outlandish and otherworldly landscapes, plus a battery of instruments capable of analysing the surfaces of the moons and telling us what they are made of. Bewitched by the tantalizing images sent back by the Voyagers, scientists were desperate to know more. Was Europa covered solely in pure water ice? How do Io's volcanoes work? What about Jupiter itself? Do its clouds contain complex organic mixtures, or mainly the star-stuff – the primeval mixture of hydrogen and helium, together with noxious clouds of ammonia, that astronomers first detected from Earth in the 1930s? As well as withstanding all the rigours of the trip, and the radiation and the cold, the proposed Jupiter entry probe would have to survive being plunged into the planet's soupy atmosphere at over 210,000 kph. It would have to avoid burning up like a meteor in the process, slowing instead to a sedate speed, then drifting down on a parachute while sending back reams of priceless information about the inside of a gas giant for as long as possible, before eventually being crushed and fried as the pressure and temperature increased relentlessly the further it descended.

Once the science community learned that JPL was planning to send a spacecraft to orbit Jupiter and take close-up photographs of its moons, scientists started to plan exactly what experiments were to be carried aboard. An infrared spectrometer would be needed to carry out chemical analysis of the surfaces of Jupiter's moons and of the cloud-tops of Jupiter itself. The swirling storm of radiation and energetic particles in the vicinity of the planet – second in intensity only to those emanating from the Sun itself – would need examining.

Even by the late 1970s there was no consensus on what the Jupiter mission should look like. The proposal was to build a double mission: a spacecraft that would orbit Jupiter, sweeping past its moons, and another vehicle – the atmospheric probe – that would parachute into the clouds of the giant planet itself, taking measurements from one of the most inaccessible places in the Solar System and somehow beaming them back to Earth. That much was agreed, but there remained the all-important question of the launch vehicle. Would it be blasted into space aboard an expendable rocket such as the Titan launcher? Or would it go up on the new Space Shuttle, a vehicle that hadn't even flown yet? The former was the approach favoured by the JPL mission design team. The Lockheed Titan was reliable and, most importantly of all, it was an existing system. The Shuttle, though by now far more than a gleam in a rocket-scientist's eye, was still at the testing stage. There were simply too many variables in designing a mission around this untried and untested technology.

One of the earliest movers and shakers on the Galileo mission was Norm Haynes. In the late 1970s he was manager of Science and Mission Design. In those days the Galileo team, under the leadership of John Casani, was tiny – just a few scientists, headed by Torrence Johnson, and engineers such as Matt Landano, trying to sketch a plan of a mission they had no real idea how it would turn out. Haynes says that, though building a mission to Jupiter was a challenge, the basic physics was already well known. "Going into orbit around Jupiter was not a big deal. We knew it could be done in theory. The reason we didn't do it earlier is because we didn't have launch vehicles big enough to get a spacecraft out there and get it into orbit. Neither did we have any way of powering a spacecraft in the outer reaches of the Solar System, where you can't use solar panels."

Blueprint for exploration

In the very early days there were two different views of what the Jupiter probe should look like. One concept, the Pioneer Jupiter Orbiter, was a spinning spacecraft that would carry a probe to penetrate Jupiter's atmosphere. This is a simple and relatively cheap way of exploring Jupiter. Because the whole probe would be spinning there was no point in installing any kind of imaging system. The Pioneer Jupiter Orbiter, which would have been built at Ames, would be a "fields and particles" probe only; that is, it would study the magnetic and radiation fields in the vicinity of Jupiter (which can be done more efficiently from a spinning spacecraft) and measure the magnetic fields of the moons (if there were any to be measured).

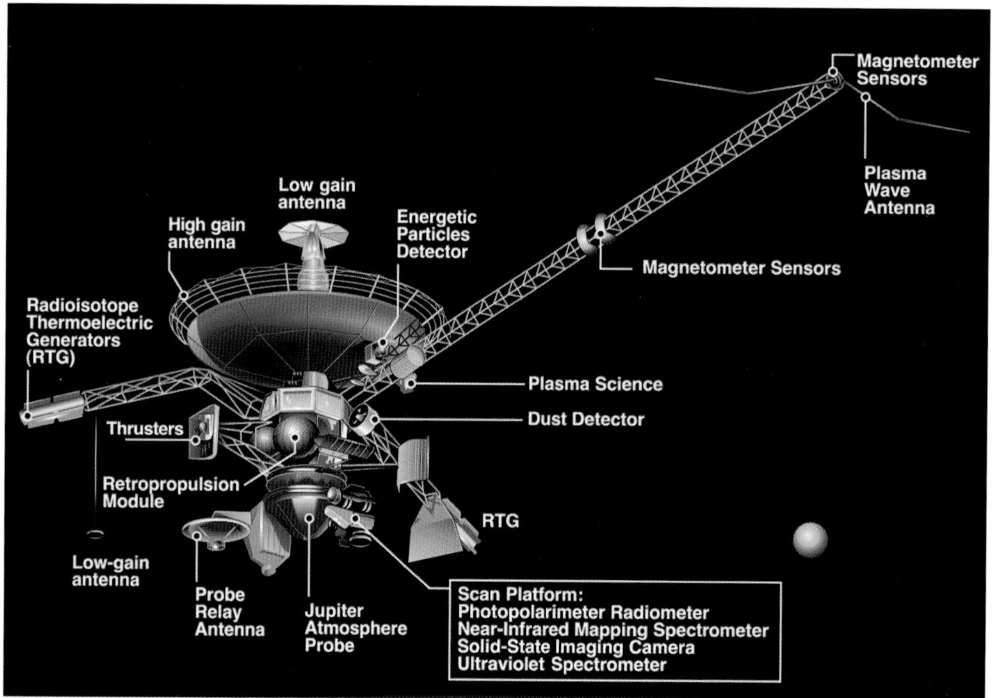

A graphic showing the components of the Galileo spacecraft.

The politicians hated this idea. Haynes recalls, "The Science Congressional Committee said, 'You've got this spinning spacecraft with magnetometers and stuff on it but that ain't gonna sell to the public, no one's going to get excited by that. So you've got to put a camera on this thing.' So we added a rudimentary imaging system at the last minute. Trouble is, putting a camera on a spinning spacecraft is not a great idea, for obvious reasons. You aren't exactly going to find it easy to take photographs."

So, back to the drawing board. The politicians said there must be pictures, so pictures there would be – after all, the twin Voyager probes were both equipped with state-of-the-art cameras. But the spacecraft also had to spin. Spinning is important for two reasons. First, it induces directional stability – any spinning object tends to keep on pointing in the same direction. Second, many of the experiments on board, notably the magnetometer, would work far better if the spacecraft turned. This instrument works much better if it can sweep through space to make accurate measurements of the magnetic field in which it is immersed. So JPL hit upon the solution of a "dual-spin" spacecraft. The bulk of the orbiter would turn, at a stately three revolutions per minute, and a small section – a scan platform, on which the cameras would be mounted – would remain stationary.

Haynes remembers, "JPL basically had to redesign the whole concept of the spacecraft and make it into what it is today, which is a very complex dual-spin spacecraft. We could not get rid of the spinning part of the spacecraft, because of the fields and particles experiments. Van Allen had pushed this thing through, and he demanded a good set of fields and particles experiments, and that meant a spinning spacecraft. Then the Science Committee wanted a camera, because they felt they needed it to keep the public on board. And so what we designed was a big spacecraft that was both. The whole thing spins, but when you wanted to take pictures you'd de-spin the back half, the bit with the cameras on, and point it to where you want to take pictures. I mean, it's the most complex spacecraft design, we, I believe, have ever attempted."

Slowly but surely, the Galileo project was taking shape. The basic design was for a large orbiter craft, weighing anything up to 3 tonnes. Attached to it would be a 1.3-metre wide, 1-metre long probe, shaped like a top and packed with instruments, that would be ejected from the mother-ship a few months before Galileo arrived at Jupiter, and make its own way to the giant planet on a slightly different trajectory.

Shuttle Diplomacy

From the initial concept, it took around 15 years before Galileo was launched. The final design was a triumph of engineering and expediency. Depending on who you believe, the craft cost anything from $1.5bn to $2bn, and at lift-off weighed nearly 2.5 tonnes – by far the largest and most ambitious unmanned spacecraft ever built, and still the most impressive machine to have completed a planetary mission. But before Galileo was launched, on 18 October 1989, the spacecraft and the team that designed it went through the wars.

Everyone involved in the early planning and design stages of Galileo assumed that the probe would be blasted into space aboard a good-old-fashioned Titan rocket, or something very similar. Titans were powerful and reliable, and had carried aloft over 150 satellites and space probes. Unfortunately, although launching Galileo atop a Titan rocket was the logical thing to do, it was not the politically correct thing to do. NASA in the 1970s was an agency bereft of purpose. Throughout the 1960s, NASA had had a goal – beat the Soviets to the Moon. But then things changed. NASA had been there, done that. Now what? The public demanded men and women in space, so they built a space station – Skylab – which suffered from a series of setbacks, mechanical failures and even an astronauts' strike, and was eventually crashed on Australia in 1980. The Russians, having abandoned their own lunar programme as a no-hoper, started concentrating on space stations and after a shaky start had real successes with the Salyut series.

NASA needed people in space to keep the public's interest. Going to Mars would be ideal – the Moonshot all over again, but much more dramatic – but no one knew if it was even possible with 1970s technology, and anyway the cost would have been prohibitive. (Current guesstimates of the bill to put a human on Mars range from $20bn to $100bn. Double the latter figure and then add some, and you would probably be near the mark). And, with a world full of insurgent reds in Asia to put down, Nixon was no space fan. The short-lived project to put a capitalist on Mars died before it could get off the drawing board. Happily, NASA had another, rather cheaper project on the drawing board that it thought would keep Congress and the public happy: a winged spaceplane that would revolutionize Earth-orbit transport, cutting by over 90 per cent the cost of sending a payload into orbit. The plan was for the first flight to be made in 1977.

The Space Shuttle was no more a reality than Galileo when it was decided that it, rather than a Titan, should send the JPL flagship on its way to Jupiter. The Shuttle had one massive advantage over the Titan – it needed people to fly it. And finding a reason to keep sending people into space was seen as vital to maintain the public's interest in the space programme. The fact that the Shuttle was an unnecessarily complicated – some would say totally unsuitable – way of carrying Galileo into space, which would have repercussions for the whole mission, was ignored in the interests of space politics.

The probe being inserted into the deceleration module.

Phil Barnett worked on the Launch Vehicle Utilization planning team for Galileo. It was his job to find the best way of getting the probe into orbit, and then on its way to Jupiter. "In an ideal world," he says, "we would not have used a Space Shuttle. We were very much wedded – culturally, intellectually, politically, institutionally and technically – to expendable launch vehicles, like the Titan. They are cheaper, period. But between 1972 and the mid-1980s, to stand there and say that the Shuttle was not cost-effective was to be un-American.

"I told everyone this up to the point where we became convinced that it was not in our best interests to tell anybody that anymore. There were tremendous political and institutional pressures to accept the Space Shuttle."The power of the astronaut corps – a powerful and influential clique in the space agency, together with a determination at the heart of government to make the Shuttle succeed – was such that by the mid-1970s NASA was actually planning to stop using expendable rockets like Titan altogether, and shut down production. Barnett continues: "The Shuttle was needed by NASA for its survival. After Apollo, NASA had no large manned space flight programme in place; they honestly believed that after Apollo they could not survive as a government institution – they needed something big. They had grandiose plans for a space station, for a lunar base and for a man on Mars. But the country was just not ready for another big expensive manned programme. We're talking Vietnam War time now, students protesting in the streets. Nobody wanted to start another big space project."

He adds: "Out of all this emerged the concept of a human piloted vehicle to get into low-Earth orbit to service the space stations, maybe to provide a base to get to Mars or a base to get to the Moon again. And through that concept emerged the Space Shuttle, but at the same time

NASA's budget was shrivelling. There was only one way that Congress was going to approve a Space Shuttle, and that was if NASA could get a lot of users on board and have them pay for flights. The only way to do that was to have a lot of flights per year and keep the cost per flight as low as possible. As incredible as it sounds, that was basically underneath the logic of getting rid of all the reliable old rockets before you even knew that the Space Shuttle was going to be successful. There were tremendous pressures, but when the time came to propose a Jupiter Orbiter Probe, we knew that we would much prefer to fly on a Titan. It was something we knew, it was there, it was available, it worked. The Shuttle was a risk."

JPL knew that the Shuttle was less than ideal, but were convinced that it could be made to work. It was all down to compromise and trade-off, but in the fraught political atmosphere of seventies NASA, that was hardly anything new. Barnett says, "There were a lot of safety requirements – because there were people on board – but in the long run I don't think that's a big challenge. I think the main reason it was bad was because it was not cost-effective. If NASA is in the game for money, the more money that goes into launch vehicles, the less money there is available for science. So from that point of view, it was an inherently a poor decision to use the Shuttle."

So JPL built Galileo to fit the Space Shuttle. The plan was to launch the spacecraft in January 1982, using the Shuttle to get it into orbit, and then, after Galileo had been lifted out of the cargo bay, to fire a powerful rocket– called an Inertial Upper Stage (IUS) – to propel Galileo from Earth orbit to Jupiter. If everything went according to plan, Galileo was to take a direct route to the outer Solar System, taking just three years to get to Jupiter and arriving in 1985.

But not everything went according to plan. Delays to the Shuttle programme pushed back the launch date to 1984. Galileo – or any planetary probe, for that matter – could not simply be launched at any time. The planets have to be in the right positions. Getting to Jupiter (unless you have the *Starship Enterprise* at your disposal) means taking off during a very limited "launch window" – one of which arrives every 13 months. And if you needed a gravitational boost from Mars along the way, the window would open up less frequently than that, for all three planets (Earth, Mars and Jupiter) would need to be in the correct relative positions. Every delay added years to the Galileo mission start date. The new launch date set for 1984 necessitated a new plan: to split the Galileo mission into an orbiter and an atmospheric probe, each launched aboard a separate Space Shuttle. Each would use a planetary IUS to get to Jupiter.

Then the plan had to be changed again. The Galileo probes were reunited, and the hope was for a 1985 launch aboard a Space Shuttle using a Centaur rocket to send the spacecraft on a direct trajectory to Jupiter. All this time, Galileo's main man at JPL, John Casani, was using a mixture of charm and persuasion to make sure Congress did not lose interest. He says, diplomatically, that the politicians were convinced of the value of the mission. "We had to try to do everything we could to keep this sense of commitment intact, and I spent a lot of time doing that."

By early 1982, the plan had changed once more. Now the idea was to use a two-stage IUS, instead of a Centaur, and use the gravity of Earth to slingshot Galileo to Jupiter. The probe would be sent on a looping trajectory that took it away from Earth, then back again, picking up speed as it swung past. Launch was now slated for 1985. Finally, in late 1982, the decision was made to launch Galileo in May 1986, using a Centaur upper stage. Once the Shuttle is in orbit, the crew lift the spacecraft and its attached rocket out of the cargo bay. After the crew have put a few kilometres between themselves and Galileo, orders are sent from Mission Control to fire the Centaur engine, and Galileo is on its way to Jupiter, four years late.

When the 1986 launch date was decided upon, Galileo had to be redesigned yet again. Engineers worked round the clock in the JPL assembly hall to get the spacecraft ready for the new configuration. Paradoxically, all these delays may very well have saved Galileo. The endless redesigns and rebuilds gave JPL's engineers a chance to perform upgrades on the spacecraft each time it was wheeled in for another refit. Modifications to the computer, in particular, were to prove their weight in gold, as better and better memory chips were plumbed into the system. Eventually, it was ready. The delicate antennas and magnetometer arm were carefully folded up, and the spacecraft was packed into a crate and loaded on a flatbed truck for the long journey from California to Florida. It may have been four years late, but the Galileo team were convinced that they were finally on their way to Jupiter.

It was a human tragedy, rather than political shenanigans, that made the *Rocky Road to Jupiter* (the title of an American TV documentary chronicling the early years of the Galileo mission) even longer – and bumpier – than JPL's worst fears. On 28 January 1986 at 11.40 a.m. a routine Shuttle launch took place. On board STS-51L *Challenger* were seven astronauts, including a schoolteacher, Christa McAuliffe, who had been selected by NASA as the first member of the public to ride into space. Even though Shuttle launches had become fairly routine (although the weekly blast-offs talked about by NASA in the programme's early days were never more than a fantasy) most people – including almost everybody at JPL – watched the launch.

Awe turned very quickly to horror on that unusually chilly Florida morning. Shortly after take-off, an O-ring, a seal on one of *Challenger*'s solid fuel boosters, ruptured, releasing hot gases which burnt a hole through into the liquid fuel tanks that supplied the main engine. This fire quickly tore the whole structure apart, breaking up the stack of rocket motors, the fuel tanks and the Shuttle itself. The fire destroyed *Challenger*, killing all seven crew. NASA was plunged into the darkest moment in its history. As the smoke cleared in the Florida skies, the whole space programme looked to be in jeopardy. The Shuttle would be grounded – for months if not years. Manned space flight itself was in doubt. It certainly looked like the end – or at least another long delay – for Galileo, which now had no way of getting to Jupiter.

Galileo design manager and mission director Neal Ausman remembers the day clearly. "I was watching the launch on TV. There were about six or seven of us that were in the room as we watched the explosion. My first thought was oh my God, what a tragedy, what a disaster. My third thought, my 23rd thought – I don't know – was, oh boy, our launch is in deep jeopardy. We certainly won't launch when we're slated to. NASA had to find out what happened with the *Challenger* and make sure that everything would be safe to resume shuttle operations, and that wasn't going to take weeks – there was no way it could take less than months, and we were not going to be able to launch in May."

In fact, the earliest opportunity for Galileo to get off the ground (assuming that the Shuttle would still be out of operation in May) was 13 months later in the summer of 1987. In the end, it was to be more than three years before the mission finally got spaceborne. The inquiry into the *Challenger* disaster uncovered a deep malaise in the manned spaceflight programme. The inquiry team, which included Nobel prize-winning physicist Richard Feynman and astronaut Neil Armstrong, found that cost-cutting, sloppiness and incompetence were the real causes of the Shuttle disaster. The Rogers Commission's 256-page report described a failure in management, which even today is used as a model in business schools of how not to make a decision. It was to be years before American astronauts flew again. Galileo was packed unceremoniously back into its crate and trundled back to California for yet another redesign.

In a new blow to Galileo, NASA bosses promptly banned the Centaur from the Shuttle cargo bay. Centaurs are loaded with liquid fuel, and the risk, however small, from a liquid-fuelled upper stage in the Shuttle's cargo bay was deemed to be too great. If the Shuttle ran into a problem on launch, and had to abort, the Centaur might explode. Also, the Centaur rocket is not a rigid structure – it is like a big pressurized balloon, requiring a complex pumping system controlled from the Shuttle to keep it rigid. The design went back to using a solid-fuelled Boeing-built IUS – a far less powerful rocket. In fact, the new two-stage IUS allowed by the launch team was so underpowered that it looked as though Galileo would not be able to get to Jupiter at all. Bill

The Galileo spacecraft takes shape inside the construction room at JPL.

O'Neil, Mission Design Manager at the time, remembers the crisis well. It was one man who managed to come up with the solution – a solution that would send Galileo not out towards Jupiter but in towards the Sun, on the start of its journey. "It was a near miracle that we managed to get there. One engineer, Roger Diehl, in a literal brainstorm, came up with idea of using the Earth and Venus to make this mission work."

Space odyssey

Galileo's thirty-month flight now became a five-year odyssey. Flying straight to Jupiter would not be possible using the asthmatic IUS, so Diehl had worked out a new trajectory which used the gravities of Earth and Venus to give Galileo the extra speed needed to whirl it out to Jupiter. Galileo would first fly to Venus, and swing by the planet, picking up speed as it did so. It would then head back to Earth, picking up more speed as it grazed past at a distance of just a few hundred kilometres. Then out to the asteroid belt, and back in to Earth again. Only after a second gravitational assist by the Earth would Galileo be going fast enough to get to Jupiter. The new trajectory was christened VEEGA – Venus–Earth–Earth Gravity Assist. VEEGA meant that the mission was still a reality, but it also meant that with a 1989 launch, Galileo would not arrive at Jupiter until late 1995, more than 10 years behind schedule. The upside was that there was yet more time to upgrade key Galileo components. Also, there was now the opportunity to do some science at Venus and during a close flyby of the Moon. The mission team may have been frustrated, but for the scientists it was beginning to look as though this little robot, which had already been through so much grief, might yet turn out to be even more of a star than was first envisaged.

Shortly before Galileo was launched, a German Earth-orbiting satellite called TVSAT was launched. It used the same thrusters used on Galileo. During TVSAT's first burn, the thrusters blew up, and an investigation showed that the Galileo thrusters had the same design flaw. JPL had to decide what to do, and quickly – change to a new set of thrusters, or make the present ones safe. The decision was to use the original German thrusters, but to pulse them, allowing them time to cool between pulses. Had Galileo not been delayed, the thrusters would almost certainly have failed at the first manoeuvre, dooming the mission.

The redesign of the spacecraft dragged on. Galileo was refitted for the solid-fuelled IUS and eventually it was packed in its crate again and driven to Florida. The launch date was set for 18 October 1989, aboard the Space Shuttle *Atlantis*. In the event, all went well, but the gods were not finished with Galileo yet. During the countdown a major earthquake rocked northern California, nearly destroying the IUS control centre. According to Neal Ausman, this was almost the last straw. "Even the Earth moved against us. It seemed like this was destined not to happen, that every time we came up with an answer, there came along a new disaster, a new problem. But we had the best, the very best people in the world. And we dealt with each and every one of these setbacks as they happened." Norm Haynes remembers: "Shuttles and then not Shuttles, and then Shuttle again, Shuttles with Centaur, Shuttles without Centaur. In a nutshell, it was awful."

At last, Galileo was on its way. The launch of *Atlantis* was flawless, and once in orbit astronaut Shannon Lucid manoeuvred the spacecraft out of the cargo bay and sent it on its way to Jupiter. As far as the men and women at JPL were concerned, the rocky road was at an end. In reality, however, it was only just beginning.

The metric system – Americans confused

NASA has had its share of foul-ups over the years – it would be impossible to do the kind of things it does without the odd mistake here and there. But perhaps the most embarrassing mistake in the agency's history came at the end of 1999, with the loss of Mars Climate Orbiter, a spacecraft dispatched to the Red Planet to survey its weather. Instead of parking itself in a nice safe orbit, as designed, the little spacecraft was on completely the wrong trajectory and plunged to its doom into the Martian atmosphere. Losing the spacecraft was a disaster for the team that built it, but even more disastrous was the humiliating revelation of what caused the accident: NASA, it seemed, had got its measurements mixed up. It emerged that a computer programmer had not realized that the spacecraft was designed and operated using metric units, and calibrated his command data in Imperial. Thus miles became kilometres and vice

FACING PAGE The successful launch of the Galileo spacecraft aboard the Shuttle *Atlantis*.
RIGHT Galileo is manouevred gingerly out of the Shuttle cargo bay.

versa, and the confused spacecraft plunged to its doom in a flurry of calibrational confusion. "NASA IN METRIC SYSTEM SCREW-UP" screamed the headlines. So, was Galileo a metric or an Imperial probe? A bit of both, it turns out. Officially, Galileo was built in millimetres and kilograms and measured its progress in kilometres, but as Neal Ausman admits it wasn't always like that. "Yes, Galileo was built in metric. But, er, well, that's not right either. We danced back and forth; we talked in Imperial and metric throughout the process. We measured thrust in newtons – but we also measured things in pounds and so on. The Americans are just not yet ready to dance with the metric system, and they didn't do a really great job when I was in JPL enforcing it."

The Galileo spacecraft

At launch the orbiter weighed nearly 3 tonnes (about the same as a medium-sized truck) including 118 kg of science instruments and nearly a tonne of rocket propellant. Galileo is huge – the graphics and pictures released by NASA do not do justice to the size of the thing. The spacecraft is as big as a small house – more than 5 m tall and equipped with a magnetometer boom that sticks 11 m out from the centre of the spacecraft.

Galileo uses very lightweight materials such as beryllium to house the subsystems, aluminium for the structure, and carbon composites for the booms that carried some of the instruments. The spacecraft comprises two main sections – a spun section and a "de-spun" scan platform attached to the bottom of the orbiter via the spin-bearing assembly (SBA), a "horribly complicated" (according to NASA) system, involving 71 metal rings and rotary transformers. The purpose of the SBA is to transfer power and data between the two parts of the orbiter. In flight, the main part of Galileo spins serenely about once every 20 seconds, although this can be speeded up to once every 6 seconds if needs be. The spinning section – the bulk of the spacecraft – contains the attitude guidance systems and the particles and fields experiments – including the magnetometer boom. The

This artist's impression shows the Galileo spacecraft, mated to its upper stage, after ejection from the Space Shuttle's cargo bay.

de-spun section consists of a ring-shaped platform housing the four instruments that need to be held steady – the cameras and other remote-sensing instruments such as the infrared mapping spectrometer, which is capable of doing "geology at a distance" and analysing rocks and surface materials from thousands of kilometres away. The design is made even more complex by the fact that Galileo's main engine – the 400-newton thruster used to slow the ship down so that it "falls" into orbit around Jupiter – pokes through the centre of the despun section, along with all its fuel lines and control wiring.

Sending data between the two sections is difficult. Obviously, because of the spinning, cables could not be used as they would wrap around one another. Instead, metal contact rings and flexible brush contacts allowed electrical messages to scuttle back and forth between the two subsections of Galileo. The design is an electrician's nightmare, and it took a long time to get right. Testing it on the ground was not easy, as the whole spacecraft needed to be suspended in the air as the two sections were spun up to speed then de-spun again. Not surprisingly, a significant number of glitches and "anomalous" signals were found emanating from the SBA during tests. In time, these were completely ironed out, and the SBA – one of the most complex and cutting-edge pieces of technology on the entire spacecraft – has turned out to be almost faultlessly reliable. Despite this, JPL's next flagship mission – Cassini, a probe currently en route to Saturn – is not a dual-spin spacecraft. Taking a "never again" attitude, JPL's engineers decided that the complexities were just too great. Norm Haynes says that even now the intricacies of Galileo have not been surpassed. "Galileo was a very complicated spacecraft at the time and required a lot of ingenuity and engineering skill to get it done. In terms of pure old engineering, it may have been JPL's finest hour."

Galileo – both parts of it – is composed of a number of subsystems together with a suite of a dozen scientific instruments. The Power Subsystem does what it says – supplies electrical power to the computers, camera and other instruments. The spacecraft uses two radioisotope thermoelectric generators (RTGs) to provide electricity. Heat from the decay of about 11 kg of plutonium-238 dioxide is converted by thermocouples into electrical current. The RTGs – a system used very successfully on the Pioneer and Voyager spacecraft – produced about 570 W at launch, decreasing at a rate of about 0.6 W a month, so the spacecraft is now operating at about 530 W. That isn't very much – "less than you need to power a hairdryer", says Neal Ausman. The fact that a dozen complex instruments, including a digital camera, two computers and a motor capable of de-spinning several hundredweight of scan platform at a respectable one revolution per 6 seconds can run on hairdryer power is an impressive feat, and a testament to the quality of the engineering.

Impressive though the RTG system is, this method of powering a spacecraft is highly controversial. When Galileo was being designed there was not as much anti-nuclear feeling as there is today; nevertheless, several environmental groups had threatened to disrupt the launch of the spacecraft, and its trips across the United States between JPL to Florida had to be under heavy armed guard – on one trip protesters actually fired at Galileo using high-velocity rifles. (It is not clear how this was supposed to minimize the risk of plutonium being released into the environment.) Environmentalists objected not so much to the use of nuclear materials *per se* – certainly not in deep space – but the small but quantifiable risk that, should something happen to Galileo at launch, for example a rerun of the *Challenger* disaster, then plutonium from the RTGs could end up being blasted all over the Florida coast, giving thousands of people a potentially fatal radiation dose. Added to that, Galileo was scheduled to return to Earth twice on its long journey to Jupiter, with a small but significant further risk of disaster. So the Greens sought (unsuccessfully) a court injunction before Galileo's launch.

Each RTG contains eighteen separate heat source modules, each housing four pellets of plutonium-238. The modules were designed to survive just about any possible mishap: a full-scale explosion of the Shuttle, an onboard fire, and re-entry into the atmosphere followed by land or water impact. The plutonium itself is in the form of an oxide ceramic, a material highly resistant to fracturing. The radioactive material is encased in protective layers of graphite, and the whole thing is housed in a bulletproof metal casing. Mission Control could in principle get it wrong – as was shown later by the Mars Climate Orbiter fiasco – and Galileo could hit the Earth's atmosphere, scattering vaporized plutonium over a whole continent. Nevertheless, NASA managed, with the help of the great popularizer of science Carl Sagan, to convince the world that the risk was far less than one in a million.

Galileo's brains

The ship's computer is divided into two parts. The Command and Data Subsystem (CDS) is Galileo's "brain". This computer, housed deep in the main chassis of the orbiter, is feeble by today's standards – with a processing power equivalent to that of an early-eighties Apple Mac or a first-generation PC. The average electronic organizer almost certainly carries far more processing power than Galileo. The CDS has several functions. First, it implements instructions received from the ground to operate the spacecraft and gather scientific data. Second, the CDS gathers together the data, packages it in the most efficient way and sends it back to Earth. Finally, like the computer on the *Starship Enterprise*, the CDS must be alert and respond quickly to any problems with the spacecraft, its subsystems or the science packages.

Although the CDS is pretty old-fashioned by today's standards (remember, Galileo was designed before a single desktop PC was in existence, before the email system was invented, before mice, CD-ROMs and the whole panoply of the modern electronic age), in some ways it is a far superior machine to today's office computers. For one thing, it is "hardened" against Jupiter's fearsome radiation. More importantly, it is flexible. Just how flexible it had to be was not anticipated by its designers. In the end, it had to be stretched far beyond its design parameters if the mission was not to be lost. The CDS performed admirably and continues to do so today.

Galileo has a second computer on board, called the Attitude and Articulation Control Subsystem (AACS). This machine is in the driving seat, aiming the spacecraft at whatever imaging target has been determined for it by Mission Control. The AACS has proved to be as flexible as the main computer, and has taken on the task of compressing imaging data as well as steering Galileo through the complexities of its Jovian odyssey.

The CDS computer loads its data into the Data Memory Subsystem (DMS), a four-track tape recorder similar in principle to the eight-track audio systems fitted to cars in the 1970s. The DMS can hold around 120 MB of data. As far as the original mission design was concerned, the ability to store large amounts of data on tape was useful but not vital; after all, the vast majority of data would be returned immediately, in real time, back to Earth via Galileo's radio transmitter.

Under the hood

Galileo has one big engine and twelve little ones. The little engines, which are mounted in two banks of six on mounting panels on its sides, steer the spacecraft, tweaking its course and orientation with brief bursts of power. The nozzles of the 10-newton thrusters, each about the size of a firework rocket, all point in different directions, making it possible to provide thrust in any direction by using one or more thrusters in combination. These engines are not very powerful – each one would only just be able to lift a 1-kg bag of flour off the Earth's surface.

The big engine is very different. This motor, developing 400 N of thrust, has to slow the whole spacecraft down so that it can go into orbit around Jupiter and not go sailing past. If the main engine failed, the Galileo mission would have been transformed from an orbital odyssey to a brief flyby of the Jupiter system – no better than what was achieved by the Voyagers. The big engine, which is about the size of a child, is an archetypal piece of German lightweight precision engineering, made by Daimler Benz Aerospace AG. Galileo has a Mercedes engine under the hood.

Phoning home

To communicate with Earth, Galileo was given a large radio dish nearly 5 m across called the High Gain Antenna (HGA). Three other radio transceivers allowed communication between the orbiter and the atmospheric probe, and with the Earth when it was in close vicinity. The smaller, low gain antennas were also to be used when Galileo was close to the Sun, during the Venus flybys. The High Gain Antenna was designed powerful enough to send gigabytes of data back to Earth – thousands of colour pictures of Jupiter and its moons, and reams of science data. The HGA was too big to fit into the Shuttle cargo bay, so it was designed to be folded up like an umbrella during the first half of its journey. Halfway to Jupiter, instructions would be sent from Mission Control to unfurl it.

On Earth, three radio stations would keep track of Galileo, twenty-four hours a day. The Deep Space Network (DSN) consists of three sets of radio dishes in California, Spain and Australia, located roughly 120° apart on the Earth's surface. The DSN ensures that at no time is Galileo or any other NASA probe "eclipsed" by the bulk of the Earth and out of contact with Mission Control. (The probe can, of course be eclipsed by other planets or moons. When this happens, the way the signal fades and rises can be used to glean vital information about the ionospheres of the objects doing the eclipsing.)

Dressed in black

Galileo's electronics and science instruments are designed to work in the hard vacuum of interplanetary space, but without some sort of insulation it's too cold for them to operate. The spacecraft is swaddled in black and gold blankets that are carefully designed to keep Galileo's vital organs at a "comfortable" temperature. They also keep micrometeorites from smashing into the spacecraft electronics.

The black blankets, which have twenty layers, are highly efficient insulators. Although only half a centimetre thick, the insulation they provide is equivalent to about 30 cm of domestic fibreglass. The black colour comes from carbon in the outer layer, which keeps electrostatic charge from building up in one spot and shorting out the electronics inside.

Black material picks up lots of heat from the Sun and emits a great deal of infrared radiation. The gold blankets, however, don't absorb a great deal of solar heat, though they do radiate well in the infrared. This material (known as second-surface aluminized kapton) therefore does an even better job of insulation. It's not used on the entire spacecraft because it was developed after Galileo was designed and built (it is used extensively on the Cassini probe). When Galileo's flight path was changed to take advantage of a Venus gravitational assist, there was enough time to apply this material to critical areas of the spacecraft.

The science instruments

Galileo carries a suite of twelve scientific instruments. The spun section is home to eight, while the other four are mounted on the de-spun scan platform. Pictures are taken by the Solid-State Imaging (SSI) camera. Mounted on the scan platform, this 29 kg instrument takes the pretty pictures demanded by public and Congress alike, and is also a vital part of the mission. The SSI camera had a big job to do: take close-up pictures of Jupiter's moons, hopefully to a resolution of a few tens of metres per pixel. (One pixel, short for picture element, is the smallest unit of picture information that can be gleaned by the camera.)

The camera has lenses but no film. It works in an identical way to the digital cameras that today can be picked up from any electronics store. Light falls on an array of electronic silicon sensors called a charge-coupled device (CCD). The CCD sensor is shielded by a thick layer of tantalum from the harsh radiation in the Jovian environment. Any one view can be imaged through eight different filters, and the separate images can then be combined electronically on Earth to generate colour pictures. The individual images are taken at single wavelengths, and thus represent a single colour, so the process of electronically combining them is akin to the generation of a colour image by the yellow, cyan and magenta inks of a colour inkjet printer. The SSI camera, it was hoped, would take about a hundred thousand high-resolution images, to be relayed to Earth via Galileo's High Gain Antenna.

Cameras can tell you what something looks like, but not what it is made of. That is the job of the Near-Infrared Mapping Spectrometer (NIMS), which is a pioneering instrument for a space probe. The NIMS has two objectives: to look at the surfaces of the moons of

Jupiter to see what they are made of, and to study the atmosphere of Jupiter to determine such things as the characteristics of the cloud layers, the variation over time and space of the constituents of the atmosphere, and the temperature of the cloud decks.

The instrument works by utilizing one of the most useful properties of radiation ever discovered – its ability to carry out chemical analysis at a distance. We know what distant stars are made of not because we have been there, sucked out some fiery gas and analysed it in a lab, but because any substance emits and reflects radiation at particular wavelengths depending on its atomic and molecular composition. The NIMS uses infrared radiation reflected from the surfaces of Jupiter and its moons to tell what is down there. (It was spectroscopic analysis from Earth that first hinted that Europa was covered in frozen water.) This instrument, with its indium antimonide and silicon detectors, is capable of carrying out a geological survey at a resolution of just a few kilometres – crude, perhaps, compared with putting an astronaut on the ground, but an incredible feat for an unmanned orbiter.

The Photopolarimeter/Radiopolarimeter instrument is used to measure the intensity and polarization of sunlight – visible and infrared – reflected and scattered from Jupiter and its moons. This gives scientists valuable insights into the temperatures of these objects, and information about the chemicals present. Carrying out a similar job to the NIMS, but in a different part of the radiation spectrum, are two ultraviolet spectrometers. These study the atmosphere of Jupiter, but are also capable of probing the mysterious aurorae that arc above the planet's poles.

All the instruments described so far are essentially telescopes bolted onto various types of detector. They use visible light and its near-cousins, ultraviolet and infrared, to probe the visible Jovian worlds. But Galileo also carries instruments designed to probe Jupiter's invisible spheres of influence. The largest and most complex of these instruments is the magnetometer, which, as its name suggests, is an elaborate compass – a huge divining rod designed to swing through Jupiter's tangled magnetic field lines and unravel their mysteries. Jupiter's magnetosphere (the bubble of space controlled by its magnetic field) is the largest single structure in the Solar System. It is a "bubble" in the Solar Wind, tens of millions of kilometres across, and if it were visible to us it would appear in the night sky far bigger than the full Moon. Inside Jupiter's magnetosphere, the giant planet reigns supreme.

The main part of the magnetometer instrument is mounted on a 9m boom attached to the spinning section of Galileo. Planetary scientists eagerly awaited this magnetometer data. What was the nature and shape of Jupiter's magnetic field? How does it interact with the "Io torus", the doughnut-shaped loop of charged particles sharing the orbit of Jupiter's innermost moon? Do any of the satellites have magnetic fields? If so, what could this tell us about their internal structure? In the event, the magnetometer managed to make one of the most spectacular discoveries of the mission.

The Plasma Instrument detects and measures electrically charged particles in space. This instrument, together with the Plasma Wave Subsystem and Energetic Particles Detector, measures the interaction between charged particles and the magnetic fields of Jupiter and its moons.

The Dust Detector Subsystem counts the small grains of matter found in space. Dust can come from anywhere – from distant supernova explosions hundreds of light years away, from Jupiter, from the satellites or from comets. Space, although it contains these particles, can hardly be described as "dusty". The particles which the Dust Detector is designed to find range in size from a ten-millionth of a gram to a 10,000-trillionth of a gram.

Radio geology

Galileo is equipped with a sophisticated radio communications system which can be used to perform scientific experiments as well as providing a link to Earth. The Radio Science experiments include the ability to probe deep beneath the surfaces of Jupiter's moons and tell astronomers what is lurking beneath their icy, volcanic or cratered exteriors. The spacecraft's radio transmitter sends a signal at a well-known stable frequency. Any change in speed that the spacecraft experiences will cause the frequency of the radio signal received at Earth to change. The amount of change depends on the change in velocity of the spacecraft relative to Earth. When Galileo passes close to Jupiter or one of the

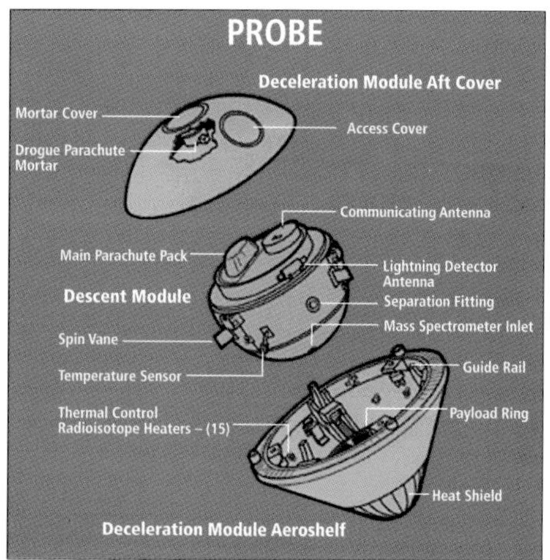

A diagram showing the Galileo atmospheric entry probe together with the deceleration module, parachute and heat shield.

big moons, that body will exert a gravitational pull on the spacecraft, causing its velocity to increase or decrease. The amount of velocity change depends not only on how big and how heavy the object is, but also how its mass is distributed within itself. Thus it is possible to use the minute – but measurable – changes in speed and direction of Galileo as it passes over, say, Europa or Ganymede to reveal something about the composition of the interior: whether the moon is layered like the Earth, with a crust, mantle and core, or is a homogenous body more like our Moon.

Finally, the Heavy Ion Counter is there as an engineering experiment rather than an astronomical observing tool. One of the things that any spacecraft has to cope with is radiation – and there are huge amounts of it in the vicinity of Jupiter. The Heavy Ion Counter measures and monitors the very high-energy ions (such as the nuclei of oxygen atoms) hitting Galileo. This information can be used to help design electronics systems – which are particularly vulnerable to ion "attack" – for future spacecraft missions.

The atmospheric probe

If fitting a dozen science instruments, two propulsion systems, two computers, four radio transmitters, fuel tanks and radioactive batteries into the Galileo orbiter was a difficult task, then building the Jupiter Atmospheric Probe was an exercise in Swiss-watch engineering. Into an object little bigger than a wheelbarrow was packed a radio transmitter, a computer to control the myriad of electronics aboard the probe, three batteries and no less than seven separate scientific instruments.

If the environment the orbiter was designed to cope with was harsh, it was a picnic compared with where the probe was going. When it slammed into Jupiter's atmosphere at a literally meteoric 210,000 kph, the outside of the 300-kg probe would be heated to 14,000°C

– two and a half times the temperature of the surface of the Sun. It would also be subjected to fierce decelerations, and further yanked about when the parachutes opened. After this extreme introduction to Jupiter, life would just keep on getting tougher for the probe as it would be simultaneously crushed and roasted by the atmosphere of the giant planet, which gets hotter, thicker and heavier the deeper you go. The probe's makers knew that eventually the titanium frame would melt and then vaporize. The challenge was to build a machine that could function for as long as possible in this harsh environment.

The result was possibly the hardiest little spacecraft ever built, a marvel of precision engineering and toughness. Before it hit Jupiter, the probe, designed at Ames Research Center and built by Lockheed Martin and the Hughes Space and Communications Company, resembled a blunt cone the size of an Irish coracle. The outside of the probe consisted of ablative heat shields – used since the early days of the space programme to protect spacecraft when re-entering the Earth's atmosphere. The materials used for the two heat shields were composite carbon fibre substances that have the properties of being extremely poor conductors of heat, protecting the precious cargo within.

Two parachutes slowed the probe on its descent. One jerked the main body of the probe from the protective descent module; the other steadied the probe as it fell gently, at a speed of just a few tens of kilometres per hour, rather than the hypersonic velocities at the point of entry into the atmosphere.

The descent module is the main part of the probe, containing the science instruments and the power and communications systems, all controlled by the probe's computer, the Command and Data Handling system. This computer was switched off when the probe was released, 147 days before the Galileo craft reached Jupiter. A timer, set to "ring" just before it hit the atmosphere, then woke it from its long slumber. Power for the computer and the science instruments came from three lithium/sulphur dioxide batteries providing about as much charge as a car battery, but weighing substantially less.

The instruments on board the probe would revolutionize our knowledge of Jupiter. The Atmospheric Structure Instrument (ASI) was the on-board weather station. Its purpose was to measure the temperature, pressure and density of the Jovian atmosphere. This machine, weighing just 4kg, had the job of analysing a gas giant. It was designed to function from the outer limits of Jupiter's atmosphere, 500 km or so above the cloud-tops, to the point where the probe inevitably fried. Unlike a terrestrial weather station, the ASI had to be capable of measuring outlandish temperatures and pressures, from 100 millibars to 28 bars – the pressure on Earth under 300 m of seawater.

The Neutral Mass Spectrometer (NMS) was to sniff the atmosphere of Jupiter and tell us what it was made of below the visible surface. It was designed to provide a detailed analysis of the chemical composition of the atmosphere and hopefully explain just what was in those swirling, multicoloured clouds. Water? Methane? Complex hydrocarbons? Sulphur? Jupiter is mostly hydrogen and helium, essentially a "raw" star, but pure hydrogen and helium would make the planet a featureless light blue. You can see that Jupiter is not blue from here on Earth, with the naked eye. Carl Sagan speculated that Jupiter was full of complex organic chemicals, amines; possibly amino acids, tars and phenols. Perhaps – just perhaps – some of these chemicals had managed to organize themselves into something even more complex – long-chained, self-replicating molecules like DNA. Could life have evolved in the swirling Jovian maelstrom?

Complementing the NMS was the nephelometer, which was to investigate the structure of clouds and the minute dust motes in the atmosphere. The nephelometer fired a laser beam from the probe through the clouds, and a reflector on an arm extended away from the probe reflected the scattered light back onto the detector. Another instrument, the helium abundance detector, measured the ratio of hydrogen to helium in Jupiter's atmosphere.

One of the oddities of Jupiter is that a world so alien should be in many ways so like home. Jupiter's weather, for instance, bears an uncanny resemblance to the Earth's. Jupiter has hurricanes, depressions, cyclones and anticyclones. Voyager also saw flashes of lightning arcing across the clouds. The lightning detector aboard the probe searched for Jovian thunderstorms. Thunder and lightning on Earth are associated with water clouds, and there is no reason to suppose that a different process is operating on Jupiter. With a bit of luck, the probe would fall straight through a thunderstorm and get pelted by rain. As well as detecting lightning, the probe was equipped with a heat radiation detector to find out which way heat flowed in Jupiter's atmosphere. On Earth, the primary source of heat is the Sun, but the atmosphere is heated by a complex interplay between incoming solar radiation, and by heat re-emitted from clouds and from the warmed ground and ocean surface. Knowing the direction of the heat flux on Jupiter, scientists could work out what powers Jupiter's weather: is it primarily a solar-driven phenomenon, as on Earth, or are those swirling hurricanes driven from below? The Doppler Wind Experiment measured the wind speed in Jupiter's atmosphere using the Doppler effect. As the probe was carried by winds during its descent, the frequency of the radio signal would change, as it is carried this way and that, just as a moving ambulance siren seems to change pitch. Voyager data showed wind speeds in excess of 300 kph near the cloud-tops. If Jupiter's weather were Sun-driven, a couple of hundred kilometres down things should be calm. The results, when they came, were quite unexpected.

After arrival at Jupiter in December 1995, the plan was for Galileo to return data from the atmospheric probe to Earth, then spend two years in orbit around Jupiter making a series of rapid passes – during orbits lasting several months – of the four main satellites (Io, Europa, Ganymede and Callisto – each pass designated with a number and a letter denoting the main target moon) and studying the Jovian environment. This was the primary mission. If Galileo were still functioning at the end of 1997, a series of extended missions would be possible. In the event, some of Galileo's most exciting discoveries would come during this extended mission.

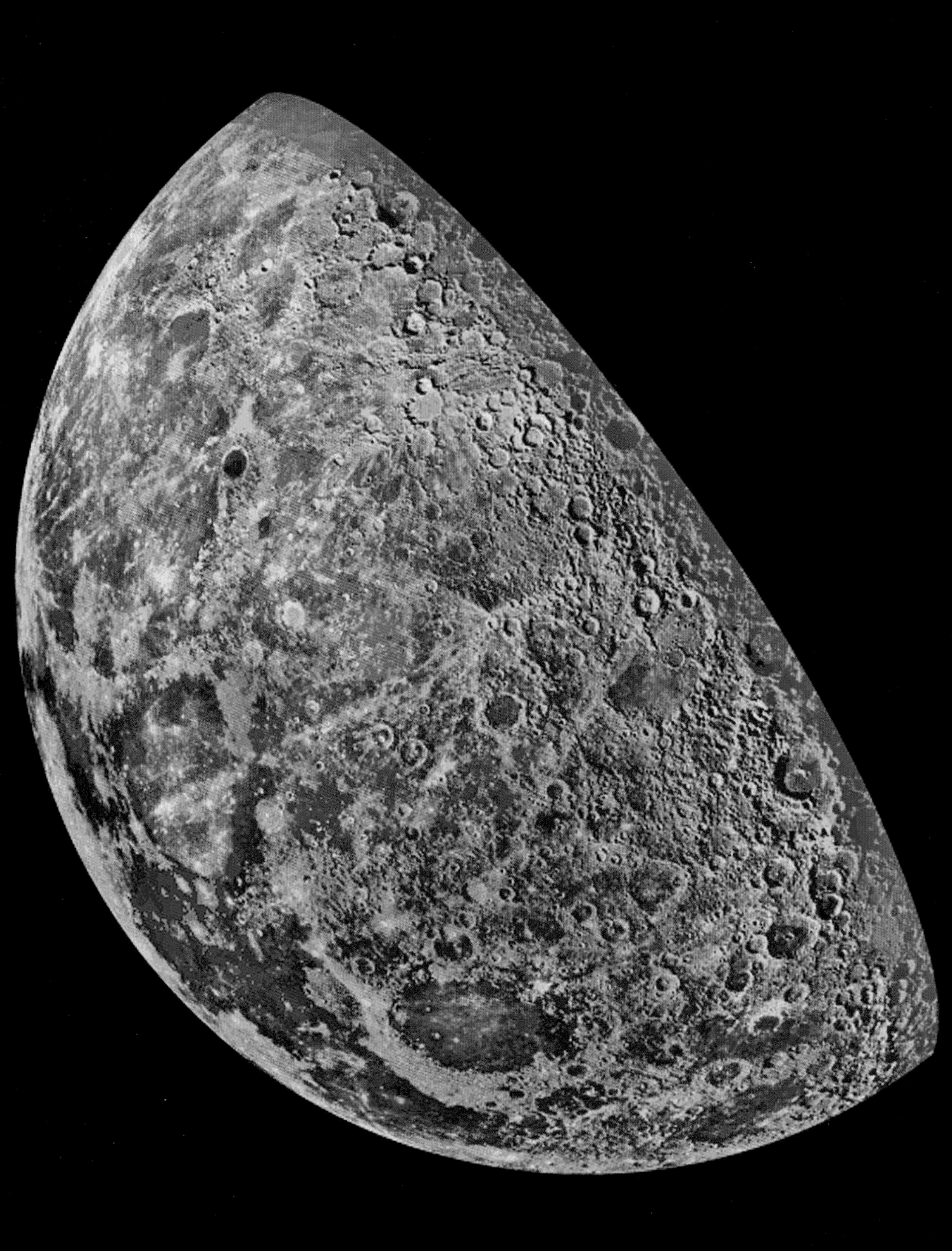

Chapter three

FLIGHT TO VENUS, RETURN TO EARTH...AND NEAR DISASTER

THE EARLY MONTHS OF Galileo's long voyage were uneventful. Just about the only action aboard the spacecraft during the four-month flight to Venus were two firings of the tiny 10-newton rocket thrusters to make slight corrections to its course. All the telemetry – the radio signals telling mission control the state of the spacecraft and its instruments – showed that Galileo was in perfect health; the endless delays, setbacks and seismic disturbances were behind it. When it arrived at Venus it was only 5 km off course, the equivalent of hitting a bull's-eye in New York with an arrow fired from London. Most importantly, it seemed as if the successive redesigns and rebuildings forced on the Galileo team as launch plans and trajectories were altered had not affected the spacecraft in any way. No one had forgotten to replace a vital widget; no one had connected wires up the wrong way. Everything appeared to be in perfect order.

On 10 February 1990 Galileo passed within 9,700 km of Venus, its first meeting with another planet and an opportunity to give the flight team a chance to open up their spacecraft and see what it could do. Peeping from

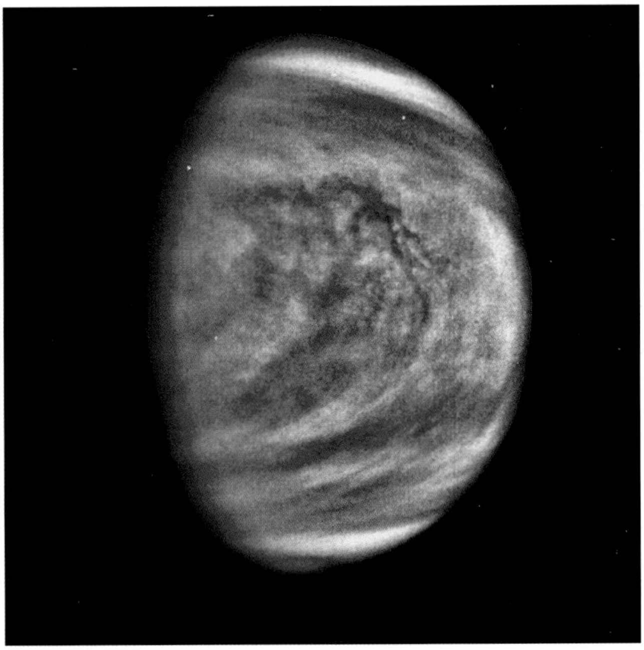

FACING PAGE The Moon in false colour. This image was taken by Galileo and approximates to a geological map of the Lunar surface. The different colours represent the different wavelengths of light and infrared radiation reflected by various rock types.
RIGHT This beautiful picture of Venus has been colour-enhanced to accentuate differences between the outermost cloud layers.

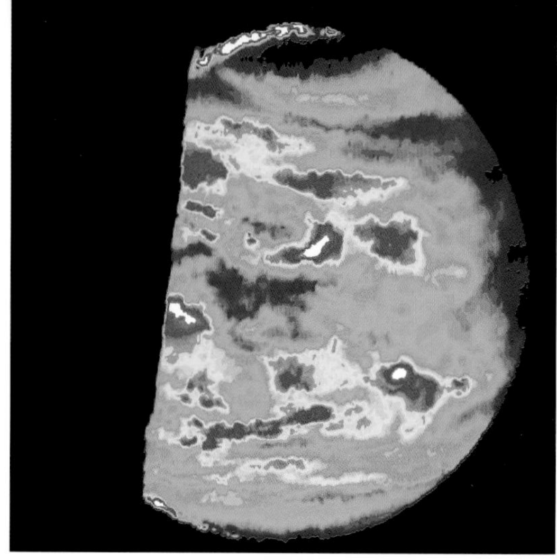

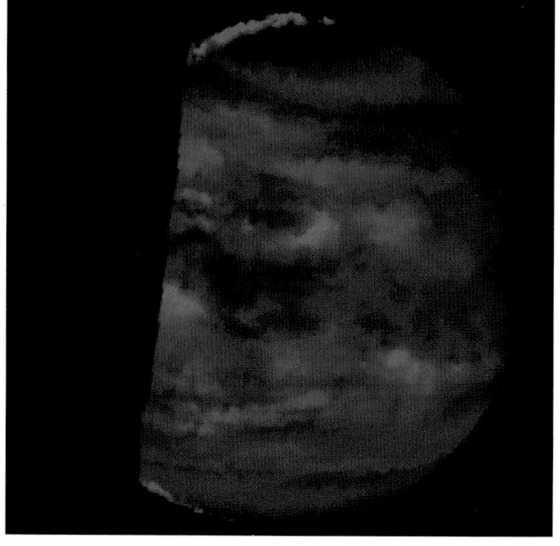

This image, in false colour, shows an infrared map of lower-level clouds on the night side of Venus, obtained by the Near Infrared Mapping Spectrometer aboard Galileo as it approached the planet on 10 February 1990. Taken from an altitude of about 100,000 km above the planet, the map shows the turbulent, cloudy middle atmosphere some 50 km above the surface, 10–15 km below the visible cloudtops. Heat from the lower atmosphere (about 200°C) shines through the sulphuric acid clouds, which appear as much as 10 times darker than the bright gaps between clouds.

behind the sunshades protecting them from the fierce solar glare, the solid-state cameras took a total of 81 photographs of Earth's sister planet over an eight-day period. Galileo is not equipped with radar, so could not peer beneath the cloud-tops of Venus, but the photographs returned were the most detailed ever taken, showing unprecedented detail of the swirling clouds at mid-latitudes. All this, remember, was a bonus. Galileo had not been designed to visit Venus at all but, remarkably, this brief encounter paid for itself a thousand times over.

Venus uncovered

Venus is a quite bizarre planet, perhaps more of a shock when it was first seen close up than any other. While few by the 1960s believed that canals and forests were to be found on Mars, many still were convinced that Venus, which was certainly warm enough for life and had a nice thick atmosphere, was a promising home for the Solar System's second ecosystem. Under her blanket of clouds, Venus, like a modest beauty, held her secrets hidden from view. Scientists not afraid of speculation, as well as science-fiction writers, took advantage of this meteorological modesty to conjure up all manner of fantasies about the charms and delights that might lay beneath. One popular image of Venus was as some sort of timewarped Carboniferous Earth, a soggy, sweltering humid atmosphere hugging mighty forests and insect-ridden marshlands. Some writers imagined Venusian dinosaurs, and the charming allegory *Perelandra* by C.S. Lewis evokes a pan-Venusian ocean of warm tropical waters on which floated gigantic rafts of vegetation the size of nations.

Ugly in close-up

Venus, it was thought, should not have been too unsuitable as an abode of life, with temperatures perhaps 10 or 15°C higher than on Earth, and rain and oceans certainly possible. Sadly, when humanity's envoys in the shape of American and Soviet robot probes finally got to the Venus in the 1960s, all these visions were shattered. Far from being a humid and sticky swamp, alive with all manner of alien pondlife, Venus turns out to be a parched oven of a world, with 450°C surface temperatures and a crushing atmospheric pressure of 90 bars. It is so hot on Venus that the rocks glow a dull red and radiate microwaves. The almost pure carbon dioxide air is so thick that a (well-insulated) bewinged astronaut could easily fly using muscle-power alone. On the highest mountains a liquid metal dew can form. The air is so soupy and has such a high refractive index that the horizon appears to curve upwards, like the world seen through a fish bowl. Hardly a breath of wind stirs at the ground; the weather is the same at the equator and at the poles. When the hardy Russian Venera probes managed to take pictures from the surface before their cameras fried (an incredible and much underrated feat of engineering), they showed a hellish world in which slabs of volcanic rock lay under a permanently yellow, lowering sky, in light levels "about the same as on a gloomy Moscow winter afternoon", according to the Russian scientists.

This is one of the most extraordinary images in the book. It shows the surface of Venus taken by the Venera spaceprobe after it had landed in October 1975. Despite being baked in the near 500°C heat, and crushed in the 90-atmosphere pressures, Venera, its camera and transmitter survived long enough to take a picture of slabby volcanic rocks beneath a lowering yellow sky.

Venus is, in short, a hellish place, certainly for humans to visit, and even our best machines have a hard time coping with the conditions on the surface (even the super-tough Veneras all gave up the ghost after an hour or so). Understanding why it is so hellish is important. On paper, Venus should be far more like the Earth than it is. It is about the same size, and was formed at the same time out of the same stuff. It has a substantial atmosphere, and while it is closer to the Sun than we are, it is right on the edge of the Sun's ecosphere, the habitable "life zone" within which the surface temperature on a solid body is conducive to the existence of life. The reason that Venus is so un-Earthlike is that carbon dioxide atmosphere: Venus, it seems, has undergone a runaway greenhouse effect that has turned the whole planet into a giant pressure cooker. Carbon dioxide is an efficient trapper of heat, allowing incoming solar radiation to penetrate to the surface, but preventing infrared radiation given out by the heated ground from leaving.

Therefore, over the aeons, Venus has heated up. Something very similar has happened on the Earth (although thankfully to a lesser extent), and the terrestrial greenhouse effect is something to which we all owe our survival. Thanks to the significant amounts of water vapour in Earth's atmosphere, temperatures at the surface are a comfortable 15°C or so. Without that blanket of water, they could be 10–20°C lower, condemning our planet to a permanent ice age. That's the upside of the greenhouse effect. Lately, some scientists have been claiming that when we pump out carbon dioxide and other gases into the Earth's atmosphere, we have added an unwanted warming factor to the Earth's climatic equation, and some have even claimed that if we continue to burn fossil fuels at the rate we do now we could trigger a runaway greenhouse effect and that would be the end of us. So, understanding Venus's atmosphere is important.

Although other spacecraft had studied Venus in more detail, Galileo was the first to come equipped with the new generation of hi-tech solid-state infrared cameras that could tease fine detail out of the superficially bland cloudscape. The photographs, some very beautiful and taken in a variety of wavelengths from the infrared to the ultraviolet, show small-scale eddies and whorls in the Venusian atmosphere that have given planetary scientists a new understanding of the weather of this odd planet. Galileo found that Venus, unlike Earth, has a slow and sedate atmosphere, with pedestrian changes in the small-scale circulation measured in days and weeks rather than hours. Unlike Earth (and Jupiter and the other gas giants, for that matter), Venus has no howling hurricanes, no tornadoes, no jet streams. Instead, the entire body of the upper atmosphere swirls slowly about the planet every four days. Down at the surface, the atmosphere – in truth more like an ocean – hardly moves at all.

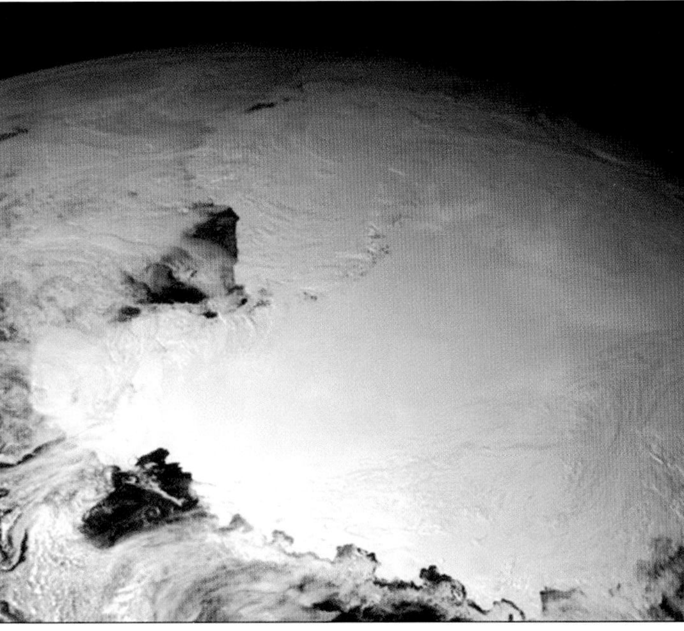

This unusual view taken by Galileo on its first pass of Earth, in December 1990, shows the continent of Antarctica, almost cloudless.

Return to Earth

Galileo picked up enough speed on its swing by Venus to head back out, away from the Sun, and in December 1990 the spacecraft made its first return to Earth for a second gravitational boost. The accuracy of its first return home was even more startling than that of the Venus approach; it arrived within half a second of its planned schedule, and was just 8 km off course. Its instruments, designed to study the alien worlds of the Jovian system, were activated and trained on the more familiar vistas of the Earth and Moon, giving the Galileo scientists an opportunity to calibrate the instruments accurately. The plasma wave experiment detected lightning during the Earth flyby, and the cameras and Near Infrared Mapping Spectrometer (NIMS) were trained on both the Earth and the Moon.

Despite having been studied for centuries from Earth, and since the 1960s by robot spacecraft and by humans directly, much about the Moon remains a mystery. By virtue of the relative orientations of Sun, Moon and spacecraft, Galileo, was able to take detailed pictures of a feature called the Orientale Basin while it was in full sunlight – the perfect conditions for spectroscopic analysis, or geology-at-a-distance.

A view of Earth taken by Galileo, centred on the Australian landmass.

Galileo found, among other things, that whatever object had ploughed into the Moon all those aeons ago to make the Orientale crater, it had not penetrated into the mantle; contrary to expectations, the so-called ejecta (rubble from the impact strewn around the crater) consisted mostly of light, crustal rocks rather than heavier materials from the Moon's deep interior. On its brief visit Galileo also managed to discover the largest crater in the Solar System – the Aitken Basin, a 2,500-km wide depression centred near the lunar south pole. The impactor that dug out this crater must have been at least 150 km across, and the impact must have shaken the whole Moon.

Galileo also trained its instruments on Earth – instruments that in most respects were far superior to those on dedicated Earth observation satellites. It was almost as though an advanced alien spacecraft has breezed past our home world. Galileo measured the hole in the ozone layer above the Antarctic, and confirmed that it was indeed growing at an alarming rate. The spacecraft's solid-state cameras took a picture of Earth every minute for 25 hours, and these 1,500 frames were run together to provide an unprecedented real-time movie of the whole Earth in rotation. As a bonus, the "movie" was centred on Antarctica, giving a unique view of our planet.

Interestingly, what Galileo did *not* detect on Earth – visually, at least – was human civilization. The cameras, though sensitive, were simply too far away to pick up any evidence of cities, roads or agriculture. However, life was detected: the bizarre composition of Earth's atmosphere, 21 per cent highly reactive oxygen, 78 per cent nitrogen and a dash of tell-tale chemicals such as methane (produced in prodigious quantities from both ends of cows) and nitrogen oxide, show that something must be alive down there. In a 1993 paper in *Nature* magazine with the tongue-in-cheek title "Search for Life on Earth", Carl Sagan gave an account of the 1990 flyby as if Galileo had visited an alien world. Working from "first principles alone" – that is, disregarding any chauvinistic assumptions about what Earthly life might be like – Sagan trawled through the data in a bid to see whether Galileo had detected life.

The Lunar farside, taken by Galileo in December 1990.

His conclusions make interesting reading: "From the Galileo flyby, an observer unfamiliar with the Earth would be able to draw the following conclusions: The planet is covered with large amounts of water present as vapour, as snow and ice, and as oceans. If any biota exists, it is plausibly water-based. There is so much O_2 in the atmosphere as to cast doubt on the proposition that ultraviolet photodissociation of water vapour and the escape of hydrogen provide an adequate source. An alternative explanation is biologically mediated photo-dissociation of water by visible light as the first step in

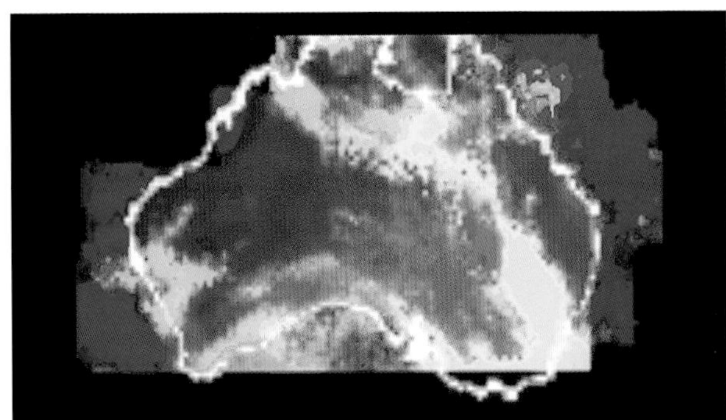

This image, taken by Galileo, shows the continent of Australia as it appears in the near-infrared.

This is one of my favourite Galileo images. It shows both the Earth and the Moon, in a "double portrait", taken from a distance of a couple of million kilometres as Galileo sped past on its final visit to its home world in 1992.

photosynthesis. An unusual red-absorbing pigment that may serve this purpose corresponding to no plausible mineral is found widely on land."

In other words, Galileo had detected a planet covered with water, in solid, gaseous and liquid forms; and it found a suspiciously high amount of oxygen, methane and large quantities of a strange green substance that may play a part in the atmosphere's unusual chemistry. If aliens had built Galileo, and if those aliens were familiar with large beasts that produce flammable hydrocarbon gases as a by-product of their digestive process, then the aliens would possibly suspect that Earth was covered with belching, farting animals. The proof of the pudding, though, came not from the visible or infrared spectra, but from the radio signals received by Galileo. "During the Galileo flyby, the plasma wave instrument detected radio signals, plausibly escaping through the nightside ionosphere from ground-based radio transmitters", Sagan's report continued. "Of all Galileo science measurements, these signals provide the only indication of intelligent, technological life on Earth."

The fact that Galileo, a sophisticated spacecraft packed with cameras and other measuring devices, was unable to tell its builders much about the complex technological civilization that spawned it, even on an almost atmosphere-shaving flyby, should alert anyone hoping to detect life on places like Mars and the moons of the gas giants. Galileo's flyby of Earth showed that finding extraterrestrial life – even if it should closely resemble life as we very much know it – will not be easy.

Disaster strikes

The success of the Galileo mission hung on the probe's ability to gather gigabytes of data while hundreds of millions of kilometres away and send it back to Earth. To achieve this, the spacecraft's designers equipped Galileo with four radio antennas – two small, low-gain antennas (LGAs) that could be used when the spacecraft was close to Earth on each of the two flybys, and also at Venus, and a large, 4.8-m flexible radio dish, the High Gain Antenna (HGA), that could be folded up like an umbrella during launch. This was the main communications antenna on the ship, Galileo's phone line to home. A fourth antenna provided a communications relay between the orbiting mother-ship and the Jupiter atmospheric entry probe.

The design chosen for the HGA was tried and trusted, and was a standard piece of kit used on many existing satellites. To the untrained eye it looks like something out of a craft shop, as if it is made of pink muslin cloth. In fact it is made of tough, gold-plated molybdenum wire mesh attached to 18 graphite epoxy ribs – the same material used in modern tennis racquets. The HGA was a key component of Galileo. Capable of sending back over 15 kB of information per second from Jupiter (a data transmission rate comparable to a slow home Internet connection), the HGA was poised to flood the Galileo science teams with data – on

Jupiter's magnetic field, the fields of its moons, the high-energy particles in the Jovian environment and, most dramatically, pictures.

Imaging chief Michael Belton and the rest of his team were eagerly awaiting the arrival of up to 100,000 high-quality images taken by Galileo's state-of-the-art digital camera. Plans were made to map large areas of Jupiter's moons and detailed weather observations of Jupiter itself. With such an impressive communications link, Galileo should even be capable of beaming home real-time moving images of the giant planet and its swirling clouds. The HGA was one of the "safest" systems on the spacecraft. Unlike, say, the solid-state camera, the large rocket engine, the computers or the dual-spin mechanism, the HGA design was certainly not innovatory. It had been tested on the ground, of course, but no one seriously thought it wasn't going to work.

The round-the-houses VEEGA trajectory that Galileo had to follow to get to Jupiter took it much closer to the Sun than was originally planned. Near Earth, an object will experience temperatures of between about -125 and +100°C, depending on its colour, reflectivity and – crucially – whether it is in shadow. (Space, being a vacuum, has no "temperature"; the temperature of an object in space depends entirely on the amount of radiation it absorbs from the Sun, and how much it radiates back into space.) The original plan, to blast off from Earth and head more or less directly to Jupiter, would have seen Galileo getting colder and colder. But the VEEGA trajectory involved a preliminary trip to Venus – about 50 million km closer to the Sun than is the Earth – where it would get considerably hotter. Galileo was redesigned to take into account these higher temperatures, and a large carbon-fibre sunshield was bolted to the top of the probe to protect its delicate instruments. The high temperatures around Venus meant that the HGA would not be deployed until Galileo was well out of the Sun's way, back beyond the orbit of the Earth.

On 11 April 1991, after the first Earth–Moon flyby, Mission Control in Pasadena sent the command for the HGA to open. A radio signal sent to Galileo ordered a small electric motor to unfurl the antenna ribs from the central mast – an action not unlike opening an umbrella. The dish, when fully open, was to lock into place. A signal would then be sent from the spacecraft telling Mission Control that all was well.

No one was seriously worried about this procedure, and 11 April was scheduled to be a pretty run-of-the-mill day.

Unfortunately, things worked out rather differently. The message was sent to release the antenna, then ... nothing. The telemetry indicated that the motor trying to push open the antenna after the pins were released was on "stall". In other words, current was flowing into the motor, but it was not able to turn. The antenna was stuck. Galileo's controllers were, to put it mildly, surprised. After all, the High Gain Antenna was rather an old-fashioned piece of space hardware, and old-fashioned, in space technology, usually means super-reliable. Neal Ausman remembers the day. "We'd used exactly the same antenna on at least three previous launches with no problems: same design. We took the spacecraft down to the Cape, installed the antenna, deployed the antenna, and there was just no question: it was gonna work. Except it didn't."

Although the failure of the HGA was potentially catastrophic, no one panicked. Galileo had a long, long way to go before it reached Jupiter, and in the interim there would be plenty of time to solve the problem. The JPL press office announced that there was an anomaly with the radio dish, but the story was more or less ignored by the media. But although Galileo's controllers were not panicking, they were not happy. If the HGA could not be made to open somehow, then the dreadful possibility loomed that even if Galileo managed to reach its destination, it would be unable to relay its findings back to Earth. No wonderful pictures of Jupiter's cloud belts, no close-up pictures of the icefields of Europa, no Ionian volcanoes in action. In short, a billion and a half dollars down the drain. Spacecraft manager Matt Landano was certainly less than pleased. "I remember we watched the currents on the motors that were driving the antenna and they almost immediately stalled. My first thought was, 'Oh my God, there's something wrong'."

His boss, Project Manager Bill O'Neil, standing next to him, was shocked. "We were gathered in the control centre, and Matt and I had tickets to Washington in our pockets – we were due to leave in a couple of hours. So, there we sat watching the screens for confirmation that the antenna was open and immediately it was obvious something

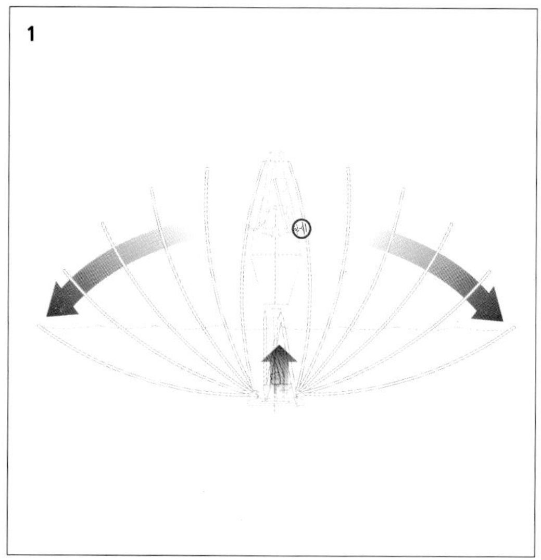

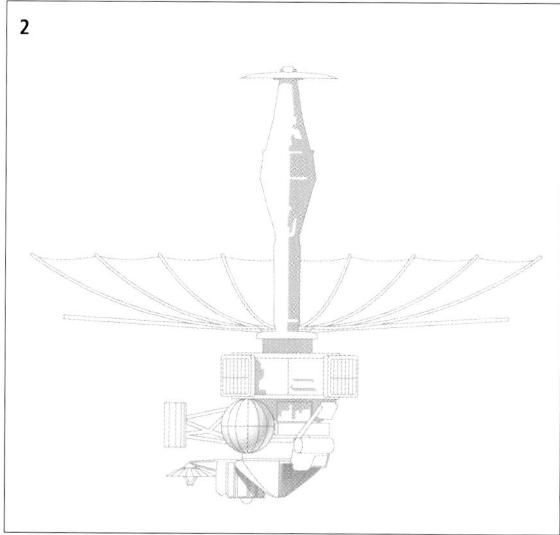

The failure of the High Gain Antenna was a near-disaster for the Galileo mission. These diagrams show the three "stuck" ribs of the antenna "umbrella", stubbornly clinging to the central tower. Diagram 1 shows the correct opening sequence; Diagram 2 shows the HGA correctly deployed on top of the spacecraft; Diagram 3 shows one of the securing pins stuck against the central tower; and Diagram 4 depicts the High Gain Antenna as it almost certainly is today, with three of its ribs stuck against the central structure. Graphics © John Lawson

was wrong – the electric currents were very high and it didn't shut off as expected. We knew we had a terrible problem right away."

Back then, everyone knew they had a problem, but no one thought it was a problem that couldn't be solved. Galileo had already brushed off the *Challenger* disaster, the wrangles over launch vehicles and even an earthquake, so a jammed antenna was not going to get in the way of the most ambitious space probe ever built by NASA. John Casani for one was not disheartened. He told O'Neil, "Boy, you dodged the bullet, it's gonna be all right, we just had this little hang-up and we'll drive it again and we'll get it over with." O'Neil ruefully admits, "Well, that wasn't the case – but that was the perception at the time." He adds, "It was perhaps less traumatic than you would think as an outsider. We were working hard and we had the confidence that we were going to get that antenna open."

O'Neil recalls, "The stall current – which meant current was feeding into the motors but they weren't turning – was an indication of a jam.

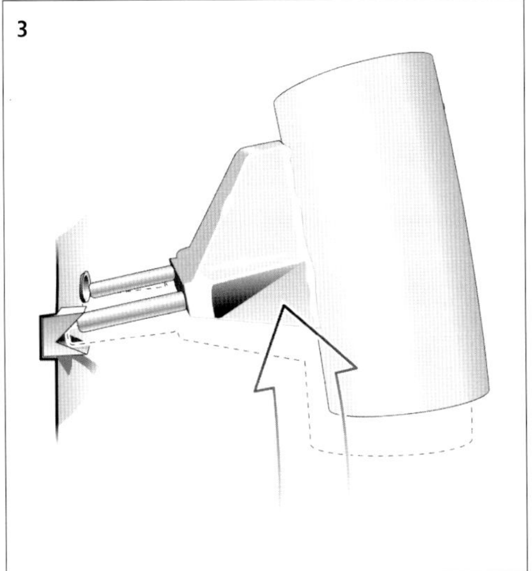

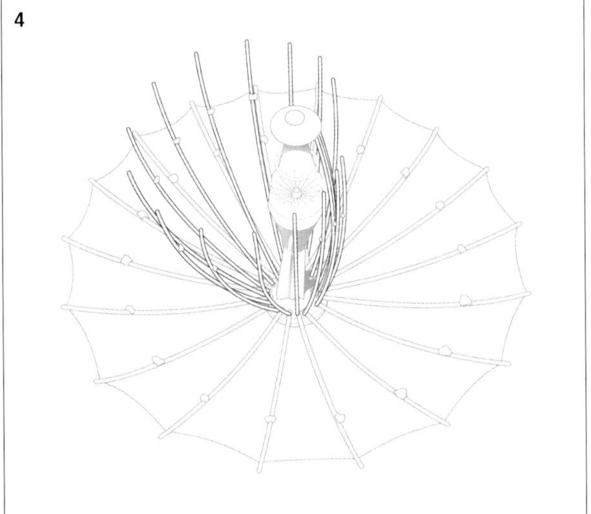

A jam someplace in the mechanisms or someplace on the antenna. We did not know where, and then we spent the next 18 months or more analysing all the telemetry and then we did a lot of ground testing with the spare antenna. We set up that spare antenna and tried to recreate what we had seen."

Eventually the specially established HGA Deployment Anomaly Team came to the conclusion that three ribs of the antenna were stuck at the central tower of the HGA, from which the ribs radiate– and furthermore that they would most likely remain that way. It was even possible to identify which of the eighteen ribs they were, as a result of looking at the altitude control telemetry and the motor currents, and the fact that at certain orientations the Sun sensor instrument appeared to have its view blocked by one of the ribs of the antenna. The wobble of the spacecraft seen after the attempt to open up the HGA also indicated that some of the ribs were properly deployed and some were not.

On the ground, test after test was run on the mock-up of the antenna. The cause of the problem was almost certainly three of the tiny metal pins designed to keep the antenna safely furled before launch. The serendipitous shading of the Sun sensor allowed the team to conclude that the problem centred on rib number 2. When a signal was sent to the spacecraft to deploy the antenna, a motor started to "wind up" the centre of the dish. When the motor started to turn, five adjacent ribs remained stuck to the central tower by their pins. As the motor continued to wind, the trapped ribs started to bend, and eventually one of them snapped free (although it did not break). Sadly, three or possibly four ribs remained stowed.

Walking round the full-scale mock-up of Galileo that is the main exhibit in the JPL museum, Matt Landano reflects wistfully on how trivial – and yet how serious – the problem was. "If I could just get out there, I could free those ribs with the flick of a finger. You wouldn't need a wrench or any tools; just a stab with a finger would do the job. You could just yank 'em out and all would be fine." The model – made from genuine Galileo parts – now has the "stuck" antenna rather than the fully deployed dish, as it was originally displayed. "We decided it would be more honest to show it how it really is", Landano says.

Why had the dish failed? The delicate-looking HGA was in fact as tough as old boots. It was built to survive the rigours of a space launch, to hold together through the shuddering 4*g* acceleration of a Shuttle lift-off. It was built to withstand the rigours of the vacuum of deep space, a vacuum that has the unfortunate effect of spot-welding metal together. It could withstand being alternately cooked and frozen, plunged from temperatures of –100°C to boiling point in just a few minutes. It could withstand the sudden vibrations and juddering of the attitude control thrusters and the big, 400-newton main engine of Galileo. What it could not survive was the appallingly dilapidated state of the US freeway system.

Road movie

In the early spring of 1986 Galileo was taken to Cape Canaveral to be launched. The spacecraft was packed carefully into a crate and placed on the back of a huge truck for its long journey across the southern United States from Pasadena to Florida. The route took in the deserts of the southwest, the swamps and bayous of the Deep South and the stifling heat of Florida itself. Matt Landano explains: "There was some talk about shipping it on an aircraft. They looked at the pros and cons of doing that, and it involved costs and risk. It was decided, since we have shipped a number of spacecraft to the Cape on a truck, that we knew how to do it. We knew the roads and we knew the people. It was simply the thing to do."

Then, after the *Challenger* disaster in April of that year (Galileo was slated for a May launch), the spacecraft had to be sent back to California for a complete redesign. The explosion of the Shuttle rightly focused NASA minds on the safety issue, and it was quickly decided that the big, liquid-propelled Centaur upper stage in a manned Shuttle was a bad idea. So Galileo had to be rebuilt to take into account that a smaller, solid-fuel IUS rocket would be used instead. This meant a new trajectory, and when VEEGA was decided upon that meant more design changes – the incorporation of the sunshield, for example. Finally, Galileo was deemed ready, and in 1989 it again began the long journey across the Deep South to the launchpad on a flatbed truck with an anxious Casani in the driver's cab. JPL now accepts that it was this long haul across the southern United States that spelt doom for the High Gain Antenna. Galileo was packed on its side in its box before being loaded onto the truck, and the constant bouncing on the potholed and rutted concrete surfaces of Interstate 10 and other roads put pressure on one side of the furled HGA. This friction caused the lubricating coating on several of the pins to wear away. "It was simply the state of the roads", says Neal Ausman.

On the long, four-year journey to Jupiter after the problem was discovered, thousands of attempts were made to free the trapped ribs. The Mission Control team tried everything. First they kept revving up the electric motor. When that didn't work, they spun the spacecraft up to maximum rotation speed – 10 rpm – in an attempt to free the stuck antenna by centrifugal force. Then they tried alternately turning the deployment motor on and off, "hammering" the mechanism in a desperate bid to free it. They tried alternately heating and cooling the dish by rotating it in and out of the Sun's shadow, in the hope that the expansion and contraction, or perhaps warming the lubrication, would somehow persuade the errant pins to work free. "The craziest suggestion", Ausman says, "came from one guy who thought we should fly Galileo into the Earth's atmosphere when it came around this way again, just clip the top of the atmosphere, and hope that the air friction did the job." As NASA had spent some considerable time, effort and money persuading an increasingly eco-conscious public that there was essentially zero chance of Galileo colliding with Earth on either of its two flybys, and discharging the deadly contents of its plutonium generators into the atmosphere, that suggestion was not taken seriously. The last official attempt to free the High Gain Antenna was made on 29 August 1994, when the deployment motor was pulsed 1,080 times, again to no effect. The control team then gave up on the HGA and concentrated their efforts on the alternative plan that was already in place.

The fact that there *was* an alternative owes a lot to the philosophy that permeated JPL in those days. Galileo simply wasn't designed to fail. Every system on the spacecraft had a back-up. If something went wrong, there was always an alternative. It was overengineered in the best possible sense of the term. Once it left Earth, Galileo was on its own: no one could come out with a tool kit to fix it. Building back-ups and back-ups of back-ups was not cheap – but then in those days "cheap" was not the byword at JPL. "If you are going to do it, you might as well do it properly" was the motto – in stark contrast to today's guiding principle of "faster, better, cheaper" that has been blamed for a series of disasters, including the highly embarrassing loss of two Mars probes in short order in 1999.

What this meant for Galileo was that all was not lost when the HGA failed. Despite being one of the most mission-critical systems, there were alternatives. For a start, there was another antenna – the much smaller low-gain transmitter designed for use when the spacecraft was near Earth (the fourth antenna, the probe relay, could not have been used to communicate with Earth). Galileo also had a powerful computer with far more memory than was actually needed. It also had a tape recorder, on which data could be stored, if necessary, before relaying it back to Earth. The team charged with the task of finding an alternative to the crippled dish had good reason to be thankful to each and every one of those back-ups, which by today's cost-cutting philosophy would never have been incorporated.

The first priority, Matt Landano says, was to make sure the atmosphere probe data could be sent back. "I remember Bill [O'Neil] asked, 'Are you sure we can do it?' We convinced ourselves that we could, because thanks to all the launch delays we had used the opportunity to beef up our computers with twice as much memory as we really needed to fly the mission." Between the original 1986 launch date and the actual blast-off in 1989, the memory chips in the CDS computers were upgraded to overcome problems that had arisen with the old type. Luckily for Galileo, as it turned out, the engineers did not have total confidence in the new chips and so packed in twice as much memory as was strictly needed: a classic case of redundancy on Galileo that was to prove its weight in launch fuel – a far more precious commodity than gold.

The problem now was that the low-gain antenna worked exactly as its name suggests: very slowly. While the HGA could have zipped back data as fast as a present-day DSL internet, the low-gain antenna was designed to transmit information at a paltry 10 bits (chunks of information) per second – about 6,000 times slower than the average home Internet connection. At that speed, a nice colour picture of the Red Spot, say, could take weeks to transmit to Earth. It was looking as though Galileo would have its performance downgraded by a factor of some 40,000. With the low-gain antenna only in operation, the Deep Space Network would struggle to pick up Galileo as it headed into the outer Solar System. Broadcasting with a hundred times less power than a mobile phone (from over half a billion kilometres away), Galileo's signal would actually be weaker than the cosmic background radiation. Clearly, a way would have to be found to speed up the low-gain system, by a factor of at least a hundred, to make the most of the weak signal. That this was eventually achieved stands as one of the greatest examples of on-the-fly space engineering in the history of planetary exploration.

O'Neill quizzed Landano. "Bill's first question was 'Can we get the probe data back?', and I said 'Yes, I believe so, but I'm gonna need a few weeks to make sure.' But I felt confident, because I designed all this. I always knew if that if the high-gain failed, we had enough gain in the low-gain antenna to play back the probe data at a very low rate."

In October 1991 a group of 36 engineers led by Leslie Deutsch and Jim Marr hatched a plan to get the data back from Jupiter. Their report, which landed on John Casani's desk on 5 November, prompted the setting up of the Antenna Recovery Mission. This team, consisting of Deutsch and Marr's original study group and 52 other engineers, began work on 9 December that year and completed their report the following March. They confirmed what Deutsch had said: the mission would work – just – using the low-gain antenna. What's more, they said they knew how to make it work.

Hero of the hour was a young programmer called Tal Brady. Just about the only person at JPL who really understood the workings of Galileo's venerable 8-bit 1802 processors, he worked night and day with his small team of programmers to rewrite the software. Jim Marr, his boss, is still amazed by the feat. "He did an amazing job. I think we almost worked Tal to death. He poured his soul into making this thing work, maybe worked as hard as anyone that I know of on the entire project, darn near around the clock, seven days a week, hardly taking time off." John Zipse, a key member of the "data systems programming team", as the Galileo rescue mission was called, died suddenly and unexpectedly during the race to save the spacecraft, but his contribution had already been made, and the others were able to carry on with his work.

Tal Brady, Jim Marr and Les Deutsch and the other 127 members of the team worked flat out from 1991 to 1995 as Galileo sped towards Jupiter. That the Orbital Tour – the bulk of the mission – could be performed using the low-gain antenna was a relief for the scientists, who were growing increasingly anxious. How on earth could they hope to explore a mini solar system with a downlink of just 10 bits per second? If the computers could not be reprogrammed in time, then the Galileo mission would be limited to just the atmospheric entry probe. A single image from the Solid-State Imaging camera of, say, Europa, would take a fortnight to download. A full survey of Jupiter's moons would be impossible. Fortunately, Deutsch and Marr were confident. A third study, led by Wayne Kohl, concluded that the atmospheric probe mission could also be saved using the low gain. By the middle of 1992, it was clear that JPL still had a viable mission to Jupiter.

"There was no way we could do the Galileo mission at 10 bits a second", Jim Marr explains. The first part of the plan was to increase

The Goldstone radio receiver/transmitter in California. Part of the Deep Space Network.

the effective downlink rate by a factor of 10 by fine-tuning both the spacecraft's communications systems and the three giant receiving antennas of the Deep Space Network A signal from a spacecraft comes in two parts – a carrier signal, and the data signal itself. The team found a way to minimize the carrier signal, so that transmissions were nearly all valuable data.

The next trick was to reprogram the spacecraft to squeeze even more into every bit sent to the ground, by compressing the data. To do this, the recovery team struck a bargain with the laws of physics and programming. They were prepared to sacrifice data quality in a big way in order to get the maximum amount of data back. In fact, the system they designed was so good that the pictures and other information returned from Galileo were probably as good as, if not better, than what would have been received had the HGA worked. This meant reprogramming Galileo's main computer – the Command and Data Subsystem. "This was primitive by today's standard, but very advanced at the time in the late seventies when Galileo was being designed", Marr says. "We designed a system that allowed the computer to take the data and store it on the tape recorder, and then we would play it off the tape recorder and feed it to the other computer, in the Attitude Control Subsystem, where it would be compressed before being beamed down to the ground."

So far, the 130 engineers and programmers had improved the low-gain system by a factor of a hundred. "We had a factor of 10 on the downlink and a factor of 10 from compression, so the strategy was not to waste a single one of those bits on the ground", Marr remembers. The feeble 10-bit-per-second performance of the low-gain antenna was supercharged into a much more respectable 1,000 bits a second. Galileo was back in the game. By using the tape recorder, it should be possible to store the science data and pictures gathered by Galileo's instruments on board, and then relay them back to Earth using the low-gain dish. Thanks to the clever programming, this could be done just rapidly enough to get the data back between each pass of Jupiter's moons during the orbital tour. It would be a tight squeeze, but they could do it. The data thus returned, even though only a small fraction of the flood of digital information that would have streamed back had the HGA worked, would allow about 70 per cent of the original science objectives to be achieved. This works because all bits aren't created equal, and some are more important than others relative to the information that you want. As project scientist Torrence Johnson says, "If you trusted it enough, a single bit would be enough to tell you whether or not there was life on Mars."

Much relief

"Everybody was pleased", says Marr. "I think the most pleased group of people were the scientists. Typically scientists are a particularly cantankerous bunch. I think everyone realized that unless everyone co-operated, there was no data coming down – no one was going to get anything out of the mission, and so the science team worked with us very, very well. Every instrument that could be reprogrammed was reprogrammed."

As well as reprogramming the telecoms systems, the science instruments on board also needed "brain transplants". The science teams, scattered across the USA and abroad, went to work. While this was going on, Galileo was taking advantage of en-route science opportunities in the asteroid belt, and would later enjoy a grandstand seat at a cosmic spectacular taking place on Jupiter itself. And all the while, the clock was ticking away. Galileo could not be parked up while essential repairs were carried out. Galileo would arrive at Jupiter, come what may, in December 1995. If it wasn't ready, the mission would be lost. Fortunately, everyone gave their all.

Jim Marr: "The spacecraft worked the first minute we turned it on. It is a very complicated process doing a complete software change-out on board. The team pretty much put their lives on hold during the time all this was developed."

The slow process of reprogramming Galileo on the fly began on 22 March 1993, almost two years after the HGA failure. The work was divided into two phases: phase one to deal with arrival at Jupiter and the atmospheric probe; phase two to cope with the orbital tour of the moons. Phase one was completed in February 1995 – four years after HGA deployment failure, and only just in time for the end of the long journey to Jupiter. The second phase of modifications was not fully completed until 1 March 1996 – well after arrival at Jupiter. It took a month to upload the new programs to the ship's computers.

The team knew that the reconfigured spacecraft would be able to achieve its main goal – the orbital tour – but only just. After arrival at Jupiter, Galileo would be put into a highly elliptical orbit around the giant planet. Only during the really close part of this orbit, where it is near Jupiter and its satellites, is the spacecraft gathering data. Close to Jupiter and its moons, the tape recorder – the "data bucket" in the words of current Galileo chief Jim Erickson, is filled. Then, when the spacecraft is in the part of the orbit when it is farther away from Jupiter, Galileo slowly empties the bucket and beams the data back to

Earth. Calculations showed that with some orbits there would be only just enough time to do this before the next encounter. It was going to be a tight squeeze.

Galileo had been given a brain transplant while 600 million km away from the surgeons performing the operation. Anyone who has ever tried to download a piece of software from the Internet will know how easy it is to get it wrong. The program may load, but it won't open; or it opens, then it crashes. This was not an option with Galileo – a "crash" could have killed off the spacecraft for good. It had to work right, and work first time. It did.

What might have been?

Despite the Stakhanovte efforts of the engineers and programmers at JPL, it is clear that the loss of the High Gain Antenna was, and remains, the major setback of the Galileo mission. The main casualty has been the pictures. Not only are pictures vital scientifically, they are desperately needed by JPL to justify its existence. In the absence of human explorers, the public has grown used to – and now demands – high-quality, full-colour images of alien worlds, and plenty of them. Had the High Gain Antenna opened, there would have been thousands of pictures of Europa, Io and the other wonders of the Jovian system. In the event, the imaging team had to be highly selective: zoom in on what looks interesting, and concentrate your efforts there. Vitally, however, the quality of what was received was excellent – probably as good as, if not better than, what would have been achieved had the HGA been operational. Neal Ausman says there was a silver lining to the loss of the HGA: it forced the science teams to be more selective and perhaps more cunning than they would otherwise have been. Nevertheless, there remains a strong sense of "what might have been" when you talk to the Galileo team.

"If we had the High Gain Antenna we would have flooded the world with pictures, maybe a hundred thousand, a hundred and fifty thousand of them", he says. "We've had to be very selective, to try and capitalize upon the limited number that we had. What we were doing is taking a little postage-stamp picture. It's like taking a close-up of a human being and you don't see the eye, you just see a little teeny tiny corner of the iris, and you say, 'What the hell is that?'" JPL insists that Galileo has been able to accomplish "around 70 per cent" of its objectives without the High Gain Antenna. This is an incredible figure, especially when you consider that had you asked any of the spacecraft's designers what could have been achieved without the HGA before it was launched you would probably have received the answer "zero".

For now, the Galileo team breathed a sigh of relief. For the next couple of years their spacecraft would have to negotiate a potentially hazardous path through the asteroid belt – the rubble-strewn zone between Mars and Jupiter. No one was sure that it would survive – even a passing blow from a rock the size of a pea would be enough to pulverize Galileo, let alone a collision with anything substantial. In the end, not only did Galileo survive its path through the asteroids, but it made one of its most spectacular discoveries in the realm of the minor planets.

Galileo was a tight fit in the Shuttle cargo bay. Thanks to the constraints imposed by the Shuttle redesign, only a relatively feeble upper-stage rocket could be used to send the spacecraft on its round-the-houses route to Jupiter.

Two decades after the first flight, the space shuttle remains Americas only manned space craft. Despite costing nearly a billion dollars to launch, political pressure has ensured that the shuttle was the first choice of launch vehicle for Galileo.

Chapter four

JOURNEY THROUGH THE ASTEROID BELT

ONE HAPPY CONSEQUENCE OF Galileo's long tour of the Solar System was that it gave the science team a good opportunity to do some sightseeing. Galileo performed some excellent work at the Earth-Moon system, and took a wonderful family portrait of our planet and its satellite to grace the walls of the JPL offices. Galileo also performed magnificently at Venus – a cold-weather spacecraft proving its worth in the warmer reaches of the inner Solar System with scarcely a glitch, save an odd incident when the camera decided to try to take some pictures without being asked.

But Galileo first started breaking new ground in the asteroid belt – it was to be the first probe to seek out and explore some of the minor planets which swarm through the space between the orbits of Mars and Jupiter. Asteroids are small objects, typically 20 km across, that litter this part of the Solar System like a rubble-strewn highway. Asteroids were once thought to be the remnants of a single planet which formed between Mars and Jupiter and then, for some unknown reason, fragmented. Now, we believe that the asteroids – there are countless thousands of them – never got the chance to form a proper planet, for the powerful gravitational influence of Jupiter simply would not allow them to coagulate into a single body. These boulders are the debris left over from the formation of the planets from the solar nebula, 4.5 billion years ago. Studying an asteroid close up would give scientists a chance to look at some of the least-understood objects in the Solar System.

There is another, more pressing need to study asteroids. Not all these flying mountains – which range in size from a few metres across to 900 km – are safely sandwiched between Mars and Jupiter. Thanks to the complex celestial mechanics of the giant planets and their gravities, the asteroid belt is subjected to all sorts of gravitational pushes and shoves that occasionally nudge an asteroid out of its orbit and send it careering

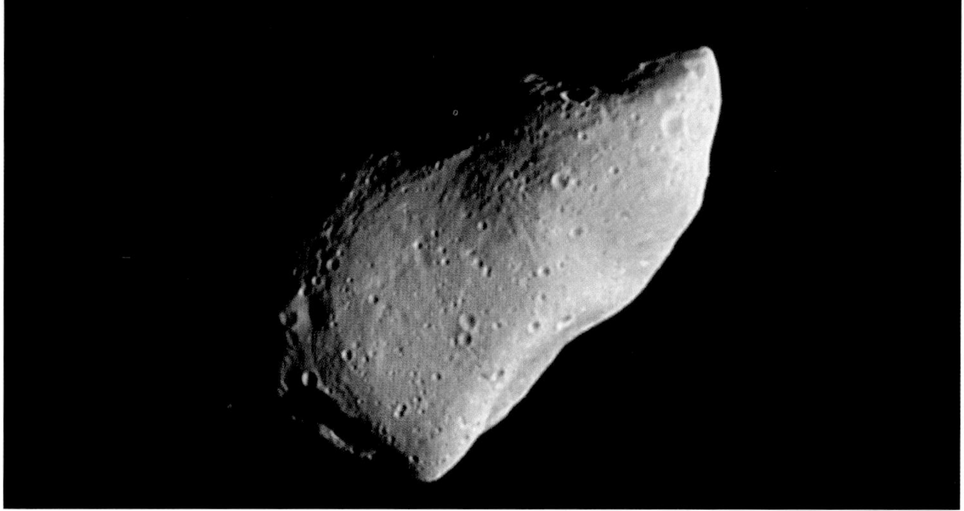

This picture of Gaspra is a mosaic of two images taken by Galileo from a range of 5,300 km, some 10 minutes before closest approach on 29 October 1991. The Sun is shining from the right. A wealth of detail can be seen, thanks to the sharp resolution of just 54 metres/pixel. More than 600 craters, 100–500 m in diameter are visible here.

sunward. Many hundreds of asteroids – some quite large – have untidy orbits that take them out beyond Jupiter or inside the orbit of Mercury. This means that occasionally the Earth is approached closely by one of these rocks, and very occasionally is hit by one of them. Even though such events are rare, they are catastrophic when they do occur: an asteroid about 15 km across ploughed into what is now the coast of Yucatán in Mexico 65 million years ago and caused havoc on a global scale. The fireball fried almost every living thing on the North American continent, and dust and ash blotted out the Sun for decades, triggering an "impact winter" that changed the climate of the Earth. Thousands of species were wiped out, including, most famously, the dinosaurs. Life on Earth has never been the same since.

More recently, on 30 June 1908, a much smaller rock fragment, or more probably a piece of comet, exploded over the skies in the remote Tunguska region of Siberia, felling trees over thousands of square kilometres. Tunguska's remoteness was fortuitous – had the cosmic interloper struck just a few hours later it would have flattened Edwardian London, killing maybe five million people. As recently as the 1930s, large meteorites capable of destroying a town fell in the South American jungle. So we have every reason to be wary of these cosmic missiles, and to find out as much as we can about them. Scientists had many questions about asteroids – were they solid bodies, or accumulations of gravel held together by their own weak gravity? The answer could someday be important for humanity – should we ever discover that one is heading our way – and sooner or later this *will* happen – then we will need to know everything we can if we are to have a chance of saving ourselves.

The Galileo science team eagerly anticipated the first asteroid encounter, with a roughly conical chunk of rock called Gaspra, on 29 October 1991. Galileo had already been to Venus, and made the first of its two return-to-Earth visits. Six months before the encounter, the High Gain Antenna had failed, so the Gaspra encounter would rely on the tape recorder to back up the data, and the low-gain antenna to transmit it (the recovery plan and the complex reprogramming operation had not yet been implemented).

As it approached the asteroid, Galileo took a series of long-range shots in the direction in which Gaspra was believed to be from ground-based observations. From Earth, asteroid tracking is a rather imprecise science: most of these objects are so small, and so dim, that knowing exactly where they are at any given time is nearly impossible. Galileo took a series of tracking images of Gaspra, honing down its position first to within the nearest 150 km, then to within 50 km.

Galileo made its closest approach to Gaspra on 29 October, when it passed the asteroid within 1,600 km. At the point of closest approach Galileo was whizzing past Gaspra so fast (at about 7.5 km a second) that the scan platform, on which the cameras were mounted, would not have been able rotate rapidly enough to track the target, so the imaging systems were only turned on a few thousand kilometres before Galileo swung past. From 16,000 km Galileo took a series of images of Gaspra using different filters, to enable a single large colour mosaic to be constructed. The non-functioning High Gain Antenna meant that no live images could be seen at Mission Control; the science team would have a long wait before the photographs could be relayed from the tape recorder. At this stage, Mission Control could not be sure that Galileo's cameras were pointed in the right direction. Just a degree or so out, and Gaspra would be out of the picture entirely.

Over the next quarter of an hour, the asteroid loomed larger and larger. When the cameras took their closest pictures, from a distance of just 5,300 km, Galileo's cameras could photograph features as small as 50 m – about the size of a small office block. Mission Control downloaded the central frame, hoping to see part of the asteroid, and got it all – the aim had been perfect. And so was the first picture. "We were overjoyed", says Imaging Team chief Michael Belton. "We'd got the asteroid right on target." When the close-up pictures were transmitted and analysed later, the scientists had their first ever chance to study an asteroid in detail.

Gaspra, which was discovered in 1916 by Ukrainian astronomer Grigori Neujmin, was classified as an S-type asteroid, meaning that spectroscopic analysis from Earth showed that it was probably made of a mixture of iron and nickel, and also dark silicate minerals identical to those found in the Earth's mantle. Galileo's photographs showed Gaspra to be an irregular, mountain-shaped object about 18 × 10 × 9 km along its three main axes. Gaspra's gravity, at just 0.05 per cent the strength of the Earth's, is so low that astronauts standing on its surface would be able to launch themselves into orbit by running or jumping. Galileo's NIMS instrument found that Gaspra, which tumbles along in its orbit like a huge spinning potato, rotates every seven hours. Gaspra bears the scars of millennia. Exposed to a slow but steady bombardment, without the protective cushion of an atmosphere, its surface is covered with craters and ridges, each testament to tremendous impacts with other asteroids and meteorites. Galileo showed that Gaspra is not uniformly grey, as had been suspected; subtle colour variations are visible on the surface. According to Joseph Veverka of Cornell University, one of the world's leading asteroid experts, Gaspra was almost certainly "derived from a larger body by catastrophic collision". Furthermore, he says, "Gaspra is a coherent body – not a pile of rubble held together by gravity."

Last visit home

In December 1992 Galileo swung by Earth for the second and last time. It needed just one more gravitational boost from our world to enable it to make the last leap to Jupiter. Already, Galileo had journeyed an incredible odyssey, hundreds of millions of kilometres this way and that, in to Venus, out to Earth, beyond Mars into the asteroid belt, then back to Earth again. It was the last time Galileo would ever be this close again to its makers. Neal Ausman says this last visit was an emotional time for the JPL team. "It was my personal farewell to Galileo. This was the last time it would come by the Earth, and then it was on its way to Jupiter. When Galileo came past there were a couple of minutes when the clouds parted and I looked outside at the Moon – the first time you could see it for days. I knew that right at that time Galileo was there, and this was a very emotional moment for me."

Astonishingly, Galileo arrived at the Earth–Moon system just a tenth of a second ahead of schedule, and less than a kilometre off target. It made more observations of the lunar geology, and dipped into a particularly close part of the Earth's radiation belts off the Argentinian coast. The flyby gave it enough of a boost to swing it right out to Jupiter. As it sped past it took a last, extraordinary picture: Earth and Moon in conjunction, a rare view of our double planet. Galileo was now on its way for real.

But not before one last close-up look at the asteroids. The Earth–Moon encounter had allowed Mission Control to tweak Galileo's trajectory slightly, using the Moon's gravity, so that it would rendezvous with a largish asteroid called Ida, a potato-shaped rock about measuring 56×25 km. As Galileo approached Ida in August 1993, its camera zoomed in on the growing object. On 28 August Galileo sped past Ida and took high-resolution colour images from a range of about 10,000 km. These images showed features just 24 m across – the size of a house. As with Gaspra, Galileo was unable to relay the Ida data to Earth straight away; it had to be stored in the tape recorder and relayed, slowly and laboriously, across the void.

Surprise

Ida turned out to be pretty much like Gaspra – a big, battered mountain of rock rich in the minerals olivine and pyroxine, as well as iron. But quite unlike Gaspra, Ida, had its own moon! On 17 February 1994, scientist Ann Harch was inspecting the first batch of picture downloads, and spotted a small object, little more than a white smudge, to the side of Ida. After checking that the object was not a star, or some other distant object aligned by chance with Ida, it was concluded that the little sphere of rock, no more than 1.5 km across – was a tiny asteroid in orbit around Ida. It was named Dactyl, after the mythical bird Zeus was reputed to have seen from Mt Ida.

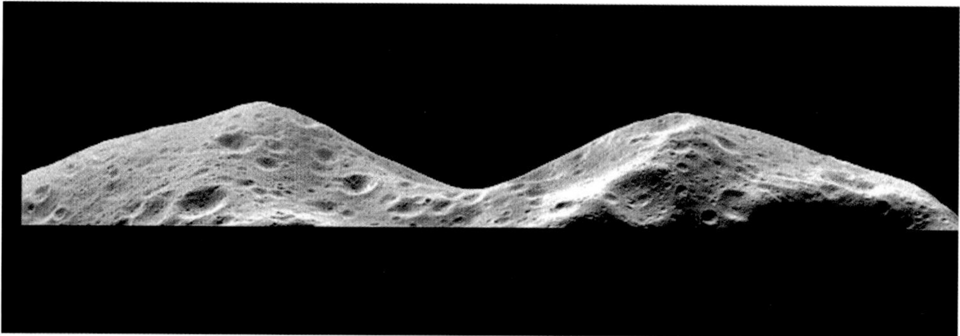

This was a lucky shot for Galileo. It shows the edge of asteroid Ida about 46 seconds after the spacecraft's closest approach on 28 August 1993, from a range of only 2,480 km. Before the successful mission to asteroid Eros by Nasa's NEAR spacecraft in 2000/01, this was the highest-ever resolution picture of an asteroid, and shows detail at a scale of about 25 m per pixel.

The Science and Sequencing Office Manager at the time was Bob Mitchell. He says that the science and imaging teams were highly cautious about their discovery. Announcing that an asteroid had its own satellite was going to make waves in the astronomical community, and they wanted to make completely sure they were not mistaken. "When we first looked at this thing, our initial thought was that it was some kind of a blemish, but we ruled that out. The next question was, is it possible that Ida has a great big crater in it and the dark that we're seeing between the rim of the asteroid and the 'satellite' is just a crater, and the satellite is actually an extended part of Ida? Well, we can pretty well rule that out just from the scale of things. So then we started looking around. Where's Jupiter at this point? Where's Saturn? Where're some other asteroids? We could have been looking at any of these."

But neither Saturn nor any other known interloper had wandered into the frame of view like an unwanted stranger in a holiday snap. Ida really did have a satellite, and the team was delighted. Mitchell recalls: "This was the first time ever there had been any incontrovertible

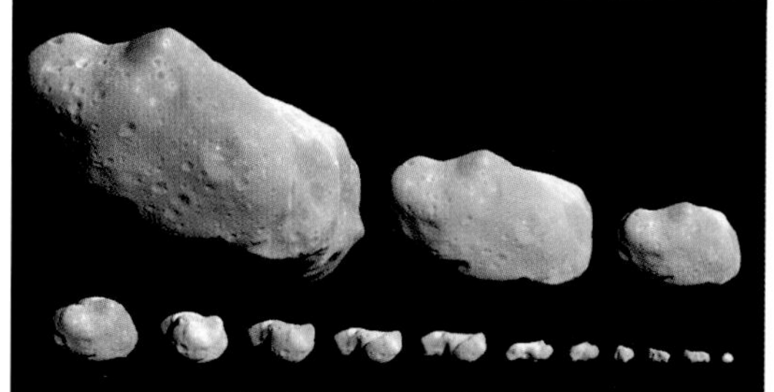

This composite image shows Ida as seen from Galileo during its approach on 28 August 1993. The six views were shuttered through the camera's green filter and show Ida's rotation over a period of about 3 hours 18 minutes. The asteroid makes a complete rotation every 4 hours 38 minutes; therefore, this set of images spans about ¾ of Ida's rotation period and shows most of the asteroid's surface.

Galileo discovered that even a little asteroid could have its own moon. This colour-enhanced image shows both asteroid 243 Ida and little Dactyl. Ida, the large object, is about 56 km long. Dactyl is the small object to the right.

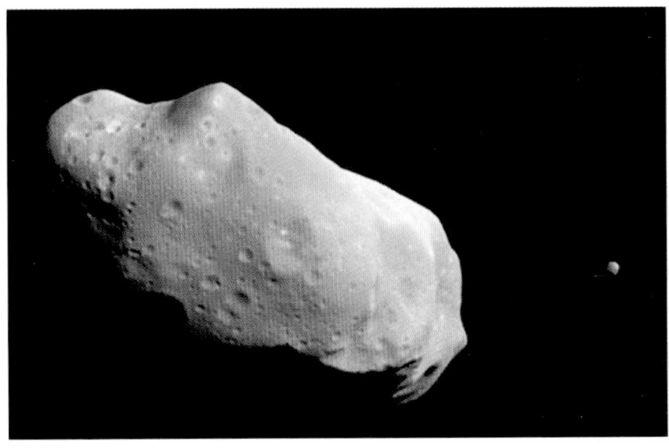

evidence that asteroids could have their own satellites. There was speculation and debate about whether that was even technically feasible, because asteroids have such small gravities. However, we proved that they can."

Galileo's visits to Gaspra and Ida had broken new ground. It was the first space probe ever to visit an asteroid and return spectacular, colour close-up images. One day the asteroids will be important to our species. As well as posing a real and terrifying threat, the asteroids are also the most valuable pieces of real estate in the Solar System. Packed with precious metals and rare minerals, the asteroids are a treasure trove of valuable materials that humankind will one day want to exploit. Paradoxically, it will be even easier to "mine" the asteroids than the much closer Moon – their tiny gravities make removing ore and sending it back to Earth, or wherever it is needed, much easier than having to blast it off the surface of a "proper" planet. Galileo has taken the first step towards exploring the Klondike of the Solar System.

On to Jupiter

After the Ida encounter – for many, one of the real highlights of the entire mission – Galileo continued on its long voyage to the giant planet. Its path through the asteroid belt was uneventful (except of course for the deliberate excitement of the two encounters). Actually, Galileo was the sixth NASA spacecraft to come this way: the two Pioneers and the two Voyagers had emerged on the far side of the belt unscathed. While the fifth probe, Ulysses, had followed a trajectory that took it well above the main swarm of asteroids. This was not unexpected; despite its popular image as a rubble-strewn band of space, the asteroid belt is far more than 99 per cent open space. The chances of hitting something were remote. As Bob Mitchell says, "There's quite a bit of material floating around in that belt, but there's also an awful lot of

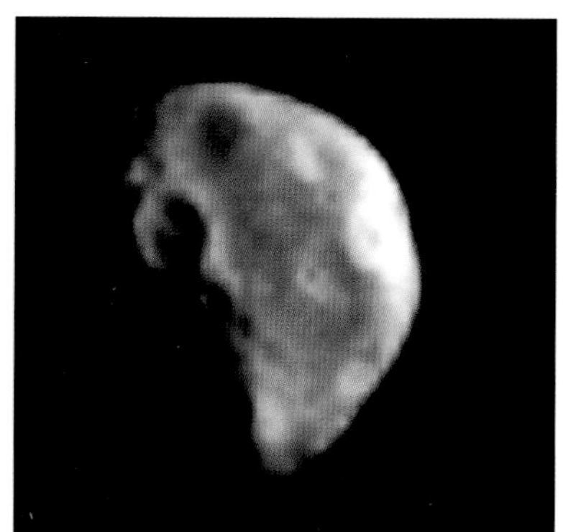

Dactyl close up! The discovery of this little world, just a couple of kilometres across, was one of the triumphs of the Galileo mission. The camera took this picture of the previously unknown moon at a range of about 3,900 km, just over 4 minutes before the spacecraft's closest approach to Ida. More than a dozen craters larger than 80 m in diameter are clearly evident, indicating that Dactyl has suffered numerous collisions from smaller solar system debris during its history. The larger crater on the terminator is about 300 m across. At the time this picture was taken, Ida was about 90 km form the moon.

space out there. So statistically your odds are pretty good. I don't know anyone who's lost any sleep over it."

Nevertheless, the Galileo team breathed a collective sigh of relief once the cosmic obstacle course had been negotiated. As Jupiter grew ever larger in its sights, Galileo had a unique chance to witness one of the most spectacular events ever seen. What Galileo saw came as a stark reminder that space is a dangerous place, and that humanity could all too easily go the way of the great reptiles, all those millennia ago.

Impact!

Husband-and-wife team Gene and Carolyn Shoemaker were veteran comet-hunters. In 1993, together with David Levy, they discovered a comet which seemed to have been ripped apart by a close encounter with Jupiter, having skimmed just 3,000 km above its cloud-tops in 1992. The gravitational trauma this had induced in the comet was enough to break it into several pieces. Comets are small balls of water ice, mixed in with exotic carbon compounds and dust, and are typically around 10 km across. They take eccentric tours of the Sun's realm. Their natural home is far beyond the orbit of Pluto, in deep space, where they pass unnoticed, black as night against the even blacker cosmic gloom. But every now and then a comet has its day in the limelight. Nudged out of hibernation by the gravity of one of the gas giants, or maybe a passing star, comets fall into the inner Solar System, where they put on a light show to rival the best that space can offer. Heated by the Sun, the volatile ices start to sublime into the vacuum of space, erupting from the comet's surface in a series of geyser-like streams of gas and dust. Lit by the Sun, these streamers can stretch for millions of kilometres, and a comet tail – which always points away from the Sun – can be the most spectacular object in the night sky.

But the comet discovered in 1993 by the Shoemakers and David Levy, using the 5m Hale Telescope on Mount Palomar, would not survive to put on a light show for the inhabitants of Earth. It had an even more spectacular fate in store. Having grazed the Jovian cloud-tops, the comet had now broken into a series of two dozen or so fragments, ripped apart by Jupiter's gravity. The fragments varied in size, with about six relatively large pieces, a dozen medium-sized ones, and assorted smaller debris, with the average chunk estimated to be about 2 km in diameter. Furthermore, the three astronomers calculated that the chunks of comet would not be able to escape the powerful pull of Jupiter. Beginning on 16 July 1994, in a firework display that would last several days, Comet Shoemaker-Levy 9 (the ninth discovery by this trio) would impact Jupiter at 200,000 kph with the force of a million hydrogen bombs. And Galileo, now 238 million km away, would have a ringside seat at the spectacle.

This was the first major impact by a comet or asteroid on a planet or moon that astronomers would ever witness, and having a keen-eyed spacecraft able to take photographs of the event was a completely unexpected bonus. As well as Galileo, astronomers trained the mighty Hubble Space Telescope on Jupiter, ready to catch the impact.

No one knew what would happen when the comet hit. Plunging into Jupiter's atmosphere at meteoric speed, the effect on even such a large planet would be devastating. The only evidence of what could happen when a comet hits a planet came from Earth. There were few eyewitness accounts of the Tunguska explosion in 1908, but the people who did see it and lived to tell the tale reported an intensely bright flash of light followed by a thunderclap from hell. The fact that no crater has ever been found at Tunguska (and remarkably little extraterrestrial debris has been recovered from around the impact site) tells us that whatever hit Siberia exploded high in the atmosphere before it had a chance to hit the ground. So astronomers knew that as the mountain-sized chunks of comet ploughed into Jupiter's thick atmosphere, there would be a series of almighty fireballs. If the comet penetrated deep, then presumably large amounts of whatever was beneath the cloud-tops would be brought to the surface, giving scientists a unique opportunity to sample the Jovian depths with Earth-based spectrometers. Some experts speculated that the impact could disrupt Jupiter on a global scale, destroying the stable cloud belts and massive hurricane systems like the Great Red Spot.

Galileo, because of its unique vantage point in space to the side of Jupiter, was able to view the direct impact sites of the comet fragments. The break-up of a comet into many fragments is an unusual event, but the collision of a large comet with a planet is an extraordinary event that probably happens only once in a millennium. As a happy JPL announced, "The fact that we had a spacecraft, Galileo, with the perfect bird's-eye view of Jupiter at the time that this happened was nothing short of lucky. Galileo had just finished its encounter with asteroid Ida when the news came in that SL9 was going to collide with

Impact from Hubble. This image of Jupiter, taken by the Hubble Space Telescope, reveals the impact sites of several fragments of Comet Shoemaker–Levy 9. The impact sites are located in Jupiter's southern hemisphere at a latitude of 44°.

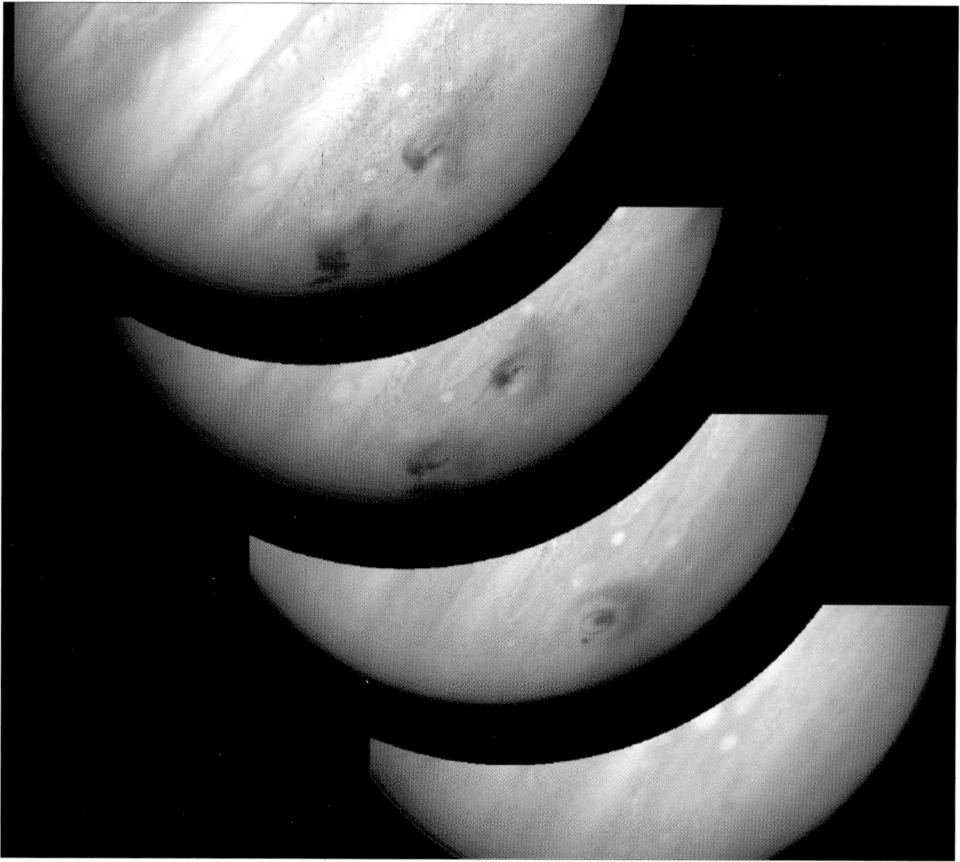

Impact Jupiter! These four, true-colour pictures, taken 2.3 seconds apart by Galileo, show the early stage of the impact of the last major fragment of Comet Shoemaker–Levy 9 into Jupiter's night side. In the first image no impact is visible. In the second picture, a bright point of light appears superimposed on the dark side of Jupiter's southern hemisphere. In the third image, the impact has grown so bright that it overexposes Galileo's digital camera. In the final Galileo image, the impact flash has faded appreciably.

Jupiter. There was a lot of controversy whether this would make a big splash or a big fizzle. Luckily for us, the science team decided to go ahead and do some observations."

The first pictures of the impact showed that the "freight train" comet had punched a series of holes in Jupiter, leaving a rash of dark scars that would take days to heal. One fragment created a 7 km fireball hotter than the surface of the Sun. After a preliminary analysis of data on fragment G (the fragments were named in alphabetical order from the first, A, to hit Jupiter) from three Galileo instruments peering from the ultraviolet into the near infrared, the project scientists had "characterized a comet impact directly for the first time in history", according to Torrence Johnson.

The impact of comet Shoemaker–Levy 9 into Jupiter was the first time in human history that people have discovered a body in the sky and been able to predict its impact on a planet in advance. If it had struck Earth, Shoemaker–Levy 9 would have killed billions of us, and possibly even wiped out our species.

This Hubble Space Telescope image of Jupiter was taken on 5 October 1995. The arrow points to the site at which the Galileo Probe plunged into Jupiter's atmosphere two months and two days later.

Chapter five

ARRIVAL AT JUPITER . . . BUT FIRST, MORE PROBLEMS

THE LONG JOURNEY FROM the asteroid belt to Jupiter gave Galileo's controllers the time they needed to upgrade the spacecraft following the failure of the High Gain Antenna, and to prepare for arrival day. During the first weeks of 1995, instructions were sent to reprogram the software to cope with the lack of a functioning HGA. The pictures from the comet impact into Jupiter were also painstakingly played back at this time. Meanwhile, back on Earth, the three receiving stations of the Deep Space Network were upgraded to make them capable of receiving the signals of almost unimaginable feebleness that Galileo would now be transmitting.

On 11 July, five months before arrival at Jupiter, preparations were made to separate the atmospheric probe from the mother-ship. Explosive bolts sheared the electronic connections between the orbiter and the squat, heavy object below it. Mission Control sent instructions to Galileo to start spinning like a top – a stately three revolutions per minute was powered up to a giddy 10 rpm. When the probe was released, on 13 July, it too was spinning, the rotation giving it a degree of gyroscopic stability. From now on, for 145 days, the probe was on its own, its computers switched off, powerless and rudderless, flying to its date with discovery – and death. The orbiter headed on, in a slightly different direction, a quick blast from the main engine tweaking its course by a fraction of a degree to ensure that, while the probe would slam into Jupiter, the orbiter would not. The release of the probe was critical to the mission. Had it not broken free from the mother-ship, not only would science have lost a priceless opportunity to sniff the atmosphere of a gas giant, but the orbiter mission would have been scuppered as well: the probe sat right behind Galileo's main engine, and had it remained in place that engine could not have been fired. Galileo would have sailed past Jupiter at 56,000 kph and on into deep space. Happily, this did not happen. "Everything is going perfectly," said Bill O'Neil at a press conference, "this is a joyous morning."

Sadly, it wasn't. All was far from well on Galileo. After wrestling with the loss of the High Gain Antenna for over a year, and coming up with an ingenious solution and implementing it, the last thing the Galileo team needed was a technical glitch, yet that is exactly what they got. And this time it looked as though the mission really might be in trouble.

On 11 October 1995, Mission Control ordered Galileo to take a picture of Jupiter, by then about 30 million km away and looming ever larger in its camera sights. The picture was to be the first test of the complex system devised by Les Deutsch and his team to rescue the mission after the failure of the HGA. First, the camera took the picture. The digitized image was stored on the ship's tape recorder, and then slowly played back to Earth via the low gain antenna. It had to work, or the mission really was lost. The JPL team was horrified when it became clear that the tape recorder had failed to stop once it had played back the image. The most plausible explanation was that the tape had snapped, and the drive spool was spinning freely. That would put the mission in serious trouble – no one could splice the two ends back together from nearly a billion kilometres away. Hearts sank further when an identical tape recorder on the ground, the one on the Galileo Test Bed in JPL, aping every move by its space-borne sister ship, tore the tape from one of the reels. By an amazing coincidence, the two machines failed within hours of each other, and from unrelated causes. Torrence Johnson was not pleased. "It's like watching one of those old movies", he said at the time. "We've had practically everything happen to this poor spacecraft on the way." Jim Marr, one of the key figures who pulled Galileo back from the brink after the HGA failure, says, "My fear was that all my team's hard work to reprogram the spacecraft would be for naught." But the Galileo mission could still be saved. Someone pointed out that a small amount of RAM (solid-state memory) in the on-board computers could be used to store science data and images – nowhere near as many images or as much data as the tape recorder could have stored, but enough to make the mission viable. Much relief all round at JPL.

The tape recorder turned out to be not as badly broken as was feared. The fact that the test-bed replica snapped its tape at just the moment when the real McCoy started playing up in space turned out to be no more than a rather spooky coincidence. It was weeks before the JPL repairmen realized that the tape had not broken, merely stuck.

Io fury

Even though the tape recorder was now known to be alive, the decision was made to scrap plans to explore Io on the approach run to Jupiter. Mission Control – in particular Galileo chief Bill O'Neil – feared that, if the tape recorder was used to store and playback Io data, in its delicate state this might be the last thing it ever did. The data from the atmospheric probe was judged to be the most important aspect of the mission – a true one-off – and this would take priority. After all, the Voyager probes had taken some pictures of the Galilean moons. Nothing had ever entered Jupiter's atmosphere, and it was unlikely that a repeat mission would be mounted within the lifetimes of many of the people working on Galileo. If the tape worked after it had returned the probe data, fine. If not, at least the priceless information from under Jupiter's cloud-tops was in the bag. This was a painful decision, as there were no plans for Galileo to revisit Io until a possible extended mission right at the tail end of the 1990s. Io was looking to be the most interesting moon at that point, and the scientists were dismayed.

It was O'Neil who had to break the bad news, especially to imaging team chief Mike Belton. This led to the most heated exchanges of the Galileo mission. A furious Belton accused O'Neil of leaving him out of the loop. He had spent years planning for the return of the Io pictures, and was now told, in a brief phone call, that he was going to go empty handed. O'Neil recalls that it was a fraught moment. "The imaging team was extremely upset that I didn't call a meeting with them to discuss the trade-offs involved in the decision. I didn't bother doing that because there was no question that we were not going to be able to use that tape recorder. Mike was very, very angry. I tried explaining to him why I did what I did, but he wasn't having any of it – he was furious that he was not included in the process. And I've never been quite forgiven for that, I hate to say. I can remember Mike's words very clearly. He said, 'Well, if it affects us next time, be sure to include us, we want to be part of the decisions and you know this has left a bad taste in our mouth.' I guess they felt bad about being left out of the decision."

An artist's impression of the Galileo spacecraft entering orbit around Jupiter. The High Gain Antenna has been shown, correctly, in its "stuck" position.

One of those most affected by the decision to "ignore" Io on arrival at Jupiter was Rosaly Lopes-Gautier, a Brazilian-American geologist in charge of the Io infrared mapping team. "I was really heartbroken", she says. "It was the right decision, as it turned out, but it was a very hard one for me to understand at the time. I had worked for a long time getting the plan for the Io observations worked out and fine-tuned. I knew these observations would be one of the major highlights of my career. My only son was born in the middle of planning for Jupiter arrival, and I stayed home only two weeks. Even then I was working and taking conference calls. To find out that all that effort had come to nothing was extremely hard."

O'Neil says that he would not have done anything differently, even if he could. "There simply wasn't time to involve everyone. This wasn't like the antenna problem – we didn't have years to work out what to do. Time was running out." In the end it was O'Neil's call, and no one challenged him again. Time has shown that he made the right decision. "It would have been so foolish not to do what we did, and it turned out, of course, that that decision was completely vindicated. We didn't have any idea at all what was wrong with the tape recorder, so we used it very gingerly to do the recording of the probe data. If we hadn't, we would have lost everything – I mean, it wouldn't only be that we had lost the probe data, we probably would have broken the recorder, so we wouldn't have been able to do the imaging of the satellites either. So it was a straightforward call."

Careful investigation using a special set of tests run on the spacecraft by a small team led by a young engineer, Greg Levanas, showed that the tape recorder was sticking on an erase head over which it needed to pass. The recorder could pull the tape off if it was going in one direction, but couldn't do it in the other; and if it stuck, the tape could end up being ripped apart, ruining the mission. Levanas and his team managed to devise a set of rules for operating the recorder which would reduce the likelihood of the tape sticking. If the Io observations had gone ahead there is a good chance that the tape would have become mangled in the machine, and that would have been the end of that. As Jim Marr says, "Ultimately a success, but heart-thumping along the way."

Arrival

Galileo entered Jupiter's domain on 26 November 1995, when it crossed the boundary of the giant planet's magnetosphere. The inbound trajectory was to take Galileo on a barnstorming close flyby of Jupiter, almost scraping its cloud-tops. During this manoeuvre Galileo would be blasted with about 40,000 rad of ionizing radiation – 40 times the dose that would kill a human being. This was to be

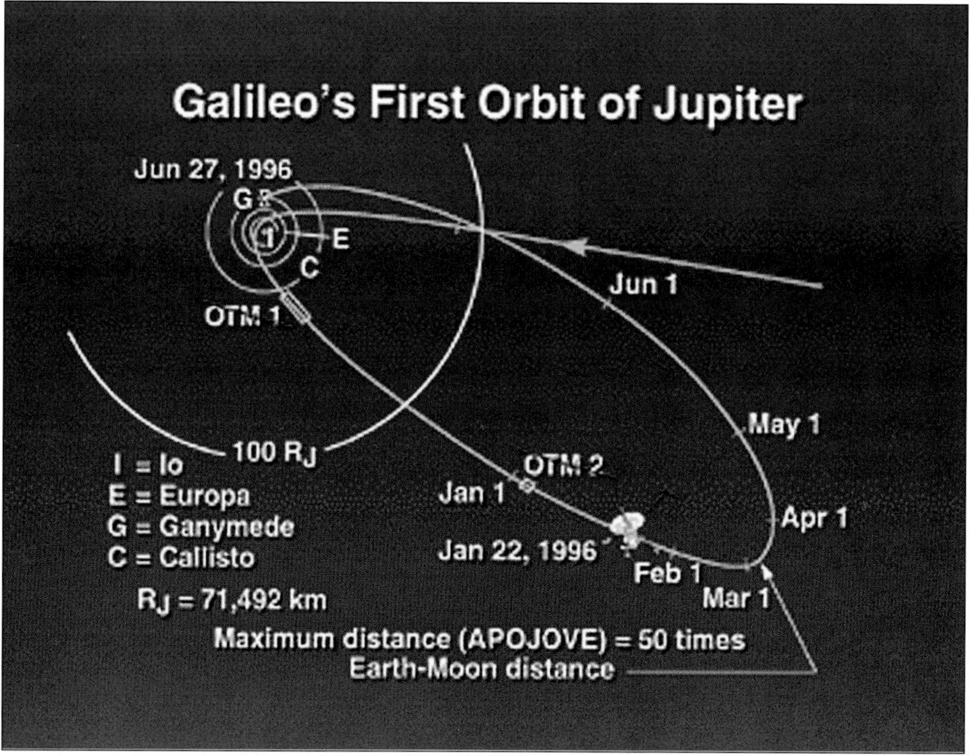

Graphic showing Galileo's approach to the giant planet, in December 1995.

about a third of the total dose it would receive during the whole of its primary mission. As senior spacecraft engineer Matt Landano said just before arrival day, "When we were building Galileo we put in a lot of shielding to offset the expected effects of Jupiter's environment, but we won't know how well a job we did until we fly through it."

Galileo had a complex piece of gravitational balletics to perform.. After swinging by Jupiter and retrieving the data from the atmospheric probe below, the main engine was to fire to slow Galileo down – a manoeuvre called Jupiter Orbit Insertion (JOI). This all had to be timed just right, otherwise Galileo would either shoot off into space, burn up in Jupiter's atmosphere or crash into one of its moons.

On 7 December 1995 the Von Karman Auditorium at JPL was packed with a thousand scientists, celebrities, reporters and cameramen. This is the room where JPL announces its big discoveries, and where it showed the first pictures to a bedazzled world. Here were shown the first pictures from the surface of Mars, sent back by the Viking landers in 1976. It was also here that the fantastic photographs taken by the Voyager probes were displayed.

Galileo had a hard day ahead of it. The spacecraft had to arrive at Jupiter, and slow down enough to fall into its gravitational well and go into orbit. Also scheduled for arrival day were flybys of two of Jupiter's moons, Io and Ganymede, transmission of the first data on Jupiter's inner magnetosphere, and a firing of the big German rocket to put the spacecraft into orbit. But the most important job for Galileo was to re-establish contact with the atmospheric probe as it plunged into the clouds, and pick up its radio signals and relay them back to Earth. All this had to be done, remember, with a bodged radio communications system.

Watching the events unfold were luminaries such as Carl Sagan, Galileo's "head of PR", now visibly succumbing to the cancer that would kill him in three years' time, and Gene Shoemaker, the discoverer of the comet that hit Jupiter (Shoemaker was killed in a car crash in Australia in 1997). Don Williams, commander of the *Atlantis* which had launched Galileo, was also present as a special guest. The cameras were trained on the screens, reporters hoping to see dramatic pictures from Jupiter. An army of press officers had to be drafted in to explain what was actually happening. No, there wouldn't be any pictures, and the events being coming through on the computer readouts actually happened hours ago, the time it takes for a radio signal to get from Jupiter to Earth.

Galileo stormed in, right on course. The short blast from the main engine three months previously had sent the spacecraft on its way with pin-point precision. This wasn't hitting a bull's eye across the Atlantic with an arrow; this was hitting a bull's eye on the Moon. Galileo skimmed by Io, darting ahead of the satellite and using its gravity to lose about 560 kph of speed. Its cameras were switched off, however, and nothing from this historic day was captured on film – which meant that elation was mixed with a degree of frustration, especially for the waiting media.

About four hours after sweeping past Io, Galileo made its closest approach to Jupiter at a distance of 214,000 km. From this here the planet would appear to an astronaut about the size of a beach ball held at arm's length – a truly awesome sight. Simultaneously, a timer aboard the atmospheric probe woke the sleeping robot from its 145-day slumber. Vital systems were switched on, computers powered up, parachutes readied for deployment. Ten minutes after the orbiter made its closest approach to the planet, the probe struck Jupiter's upper atmosphere, friction heating up its outer protective layers to thousands of degrees. Within seconds, the outside of the probe was flaming like a meteor, streaking across Jupiter's evening sky. It entered the atmosphere from the west, heading east, the sunset terminator speeding past. The point of entry was a few thousand kilometres north of the equator, on the boundary between two of Jupiter's largest coloured bands – the bright Equatorial Belt and the much darker North Equatorial Belt. Galileo's mission planners had reasoned that a cloud belt boundary might contain some of the most interesting weather on the planet, and they turned out to be right. The probe had to hit Jupiter at exactly the right angle. Just 1.4° steeper, and it would have fried to vapour before it even reached the clouds; 1.4° shallower, and it would have skimmed off Jupiter like a skipping stone and bounced away into space. In the event, it was just 0.2° out.

The probe began its long descent into the Jovian clouds. For over an hour its radio transmitter sent back megabytes of data to the main probe as it swooped overhead. There were no cameras aboard the probe, but its instruments probed, sniffed and peered into the very molecules of Jupiter's vast swirling atmosphere. The results, which are detailed in the next chapter, have revolutionized our knowledge of the giant planet.

Back in Pasadena, the tension at JPL was palpable. No one really knew if this was going to work. The probe could be a blob of glowing alloy by now, its instruments fried before getting a chance to gather precious data. The Galileo orbiter itself could fail. So much could go wrong. What everyone was waiting for was a signal that the orbiter had successfully received the data from the probe. That would mean that, basically, the mission had succeeded. In fact, as the TV cameras zoomed in on the faces of the scientists, the news everyone was waiting for had already happened. Jupiter was so far away that the signal would take nearly an hour to reach them.

Eventually, the time came for the signal to arrive. Nothing. Seconds ticked by. Still nothing. So far, despite the horrendous problems with radio dishes and tape recorders, the flight of Galileo had run like clockwork. Now it looked as if all the hard work may have been in vain. Had the probe broken up on entry? Had the radio failed? Was there an as-yet unforeseen problem with the orbiter? Could the radio signal have been pointing in the wrong direction?

The tension grew. Then a technician monitoring the transmissions screamed out,"Signal's come through!"The room erupted in applause. Torrence Johnson: "The orbiter asked the receiver aboard the spacecraft, 'Hey, have you heard from the probe yet?' And it said, 'Yep!'." Another scientist joked, "Of course, we presume it was from the probe!" Bill O'Neil watching, smiling, simply said he was "ecstatic". Dan Goldin, NASA chief, went round punching the air and shouting in his inimitable Brooklyn drawl, "Hey, is this a great day, or what?" to anyone who passed. The Galileo Website received a quarter of a million hits, something of a record at the time.

Near-disaster

But the mission wasn't out of the woods yet. An hour after the atmospheric probe "died", crushed and fried in Jupiter's atmosphere, a signal was sent to Galileo and for the second time its main engine was ignited. The orbiter was spun up to maximum revs to stabilize it, and the large engine ignited. The plan was to shave about 2,000 kph off Galileo's speed – in effect, to use the engine as a brake, pointing in the direction of travel, so that Jupiter could capture the spacecraft in its gravitational field. If the engine failed to start Galileo would sail on past the gas giant, and never have a chance to look at its moons in close-up. The probe data could still be relayed to Earth, but that would make the mission at best a very partial success. Had it misfired, Galileo could have been sent spinning helplessly into the void, like Darth Vader's fightership in *Star Wars*. Unlike Lord Vader, though, it is unlikely that Galileo would have made a return appearance. There would have been no sequel, no probe data. A 100 per cent failure.

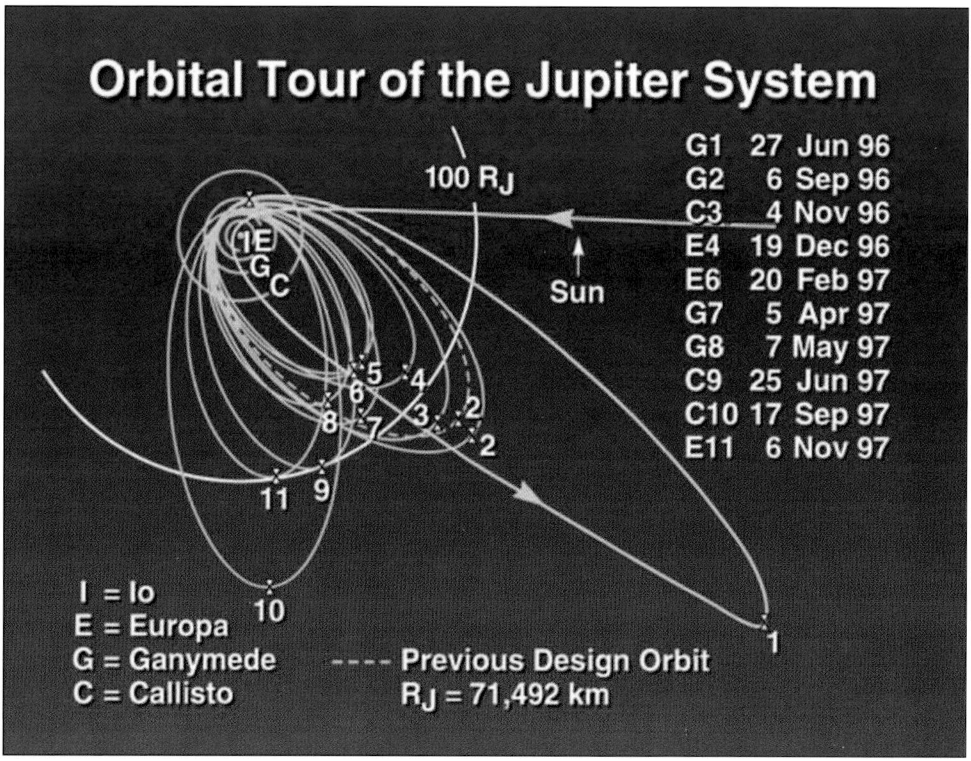

Orbital tour. This graphic shows Galileo's complex trajectory during the two years of the primary mission – which included 10 close encounters with Jupiter's moons.

At the time the world was told that everything went well. The burn itself was perfect. Once again, everyone was ecstatic. The engine fired, and Galileo slowed down and was captured by Jupiter. The firing was so spot-on that subsequent scheduled "trim burns" to tweak the trajectory proved unnecessary. Now there was a good chance that the probe's data was safe and sound aboard the Galileo orbiter, but the scientists at Pasadena and elsewhere had a wait of several days for the findings to be relayed back to Earth via the tape recorder. The relay of the data was completed just before Jupiter passed behind the Sun – blocking off all contact with Earth for a fortnight. All seemed perfect. But according to Bill O'Neil, things were not going well. "Actually, we were struggling with the threat of the propulsion system blowing up", he says.

Telemetry from the spacecraft showed that fuel from the two tanks supplying the main engine was sloshing around and stood a real chance of making contact. Galileo's engine relies on a powerful chemical reaction taking place between two different fuels. When they are allowed to mix in a controlled fashion, they ignite spontaneously (Galileo has no blue touchpaper or spark plugs) and roar out of the back in a rocket blast. A malfunction in the fuel delivery system threatened to mix the chemicals before they reached the combustion chamber, which would cause the engine – and most probably the whole spacecraft – to blow itself to smithereens. "It became a very exacting process", O'Neil says now with a degree of understatement. Happily, the moment of danger passed.

Galileo was now a prisoner of Jupiter, locked in a perpetual gravitational dance around the planet and its moons. Although its first orbit would take it millions of kilometres away from Jupiter, it would never be able to escape. The journey was over. The odyssey had just begun.

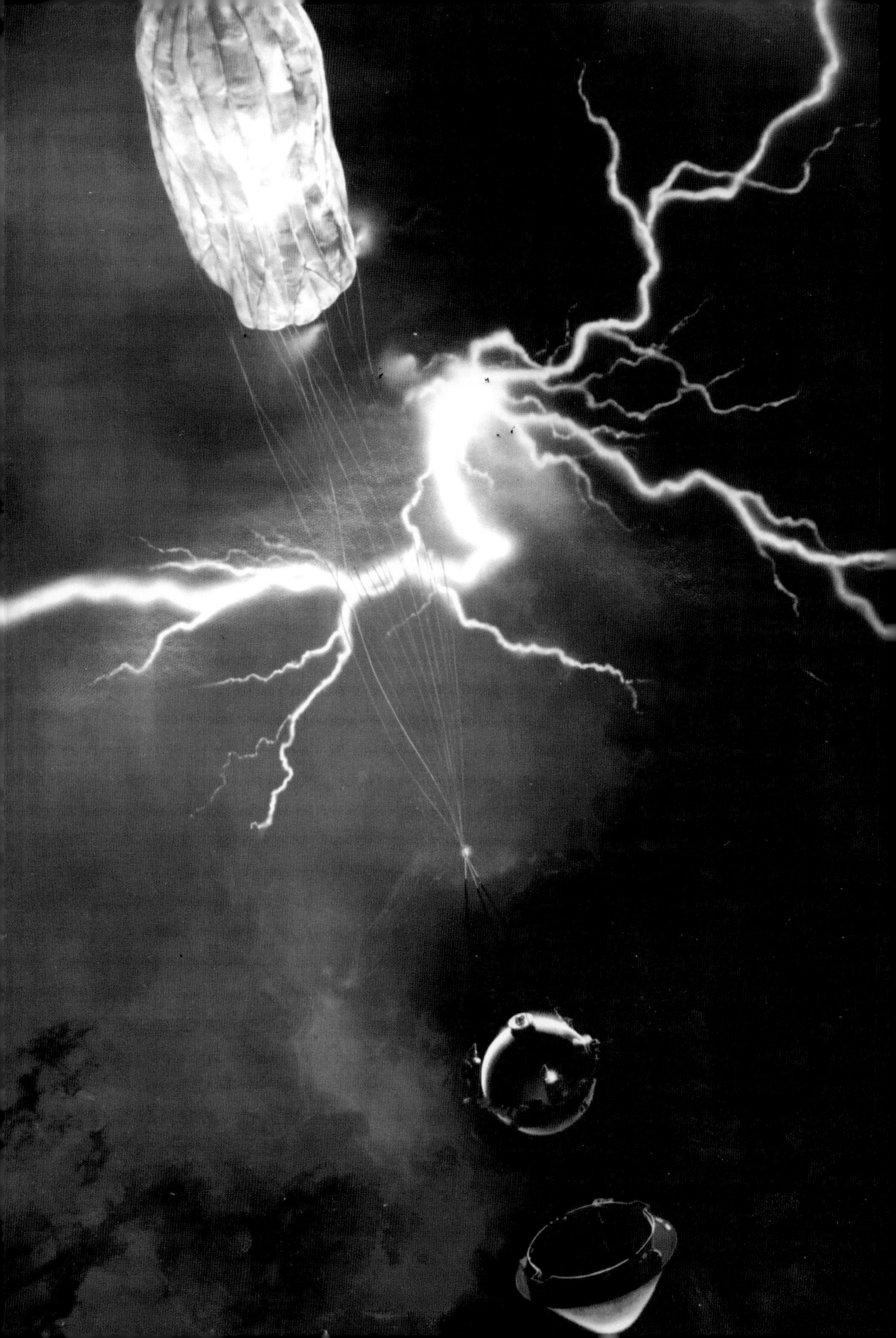

Chapter six

DESCENT INTO HELL

THE SUCCESSFUL ATTEMPT TO explore the interior of Jupiter by the Galileo atmospheric probe was one of the highlights of the mission. The technological triumph of building a machine little bigger than a wheelbarrow that not only survived being slammed into a failed star at 50 km per second, but which sent back reams of data telling us everything from atmospheric composition to light levels and a detailed weather report, was a marvel, and the probe's designers at Ames deserve the highest praise.

Many people were puzzled that the atmospheric entry probe did not carry a camera. Surely pictures from "inside" Jupiter would be among the most spectacular from the mission? Sadly, a camera would not have been practical. For a start, the extreme speeds – and particularly the extreme decelerations experienced by the probe – would have strained even the most robust of cameras. Then there was the issue of data transfer. Pictures eat up bandwidth, and sending back colour video clips from below Jupiter's cloud-tops, or even monochrome stills, would have swamped the transmission systems.

From the data returned by the probe it is clear that a hypothetical astronaut who might have accompanied the little machine on its kamikaze journey would have experienced one of the greatest shows in the Solar System. Galileo gave us a chance to visualize what such an adventurer would see on

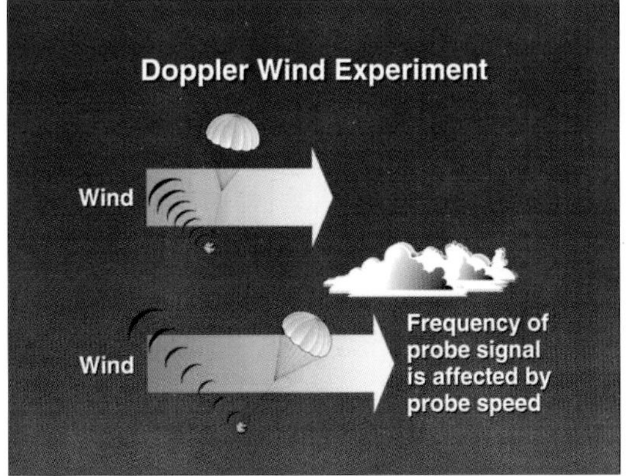

FACING PAGE An artist's impression showing the atmospheric probe descending into Jupiter's clouds.
RIGHT By tracking changes in the frequency of the probe signal, its movements through Jupiter's turbulent atmosphere could be calculated.

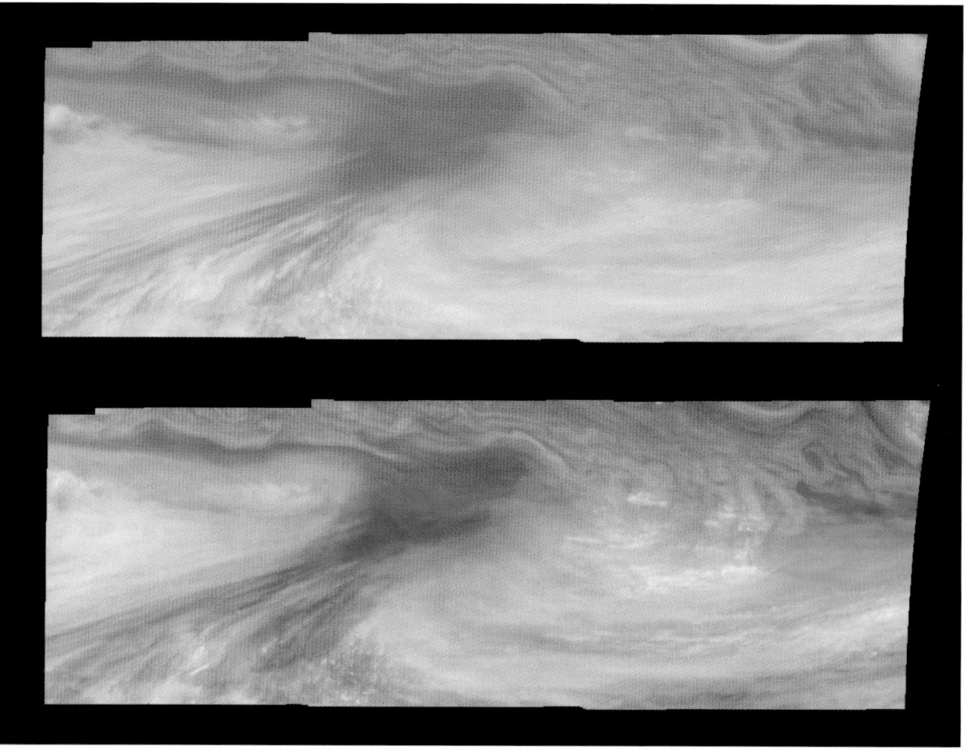

These false-colour images show a Jovian "Hot Spot", an area of intensely dry air similar to that into which the atmospheric probe plunged.

such a perilous (and certainly one-way) journey into a place that is darker and more frightening than Dante's most gruesome imaginings.

Our fictional traveller would need to be well protected as he nears Jupiter. Crossing the orbit of Io, the probe and its passenger is subjected to an incredible blast of radiation, enough to kill a human in an hour or less. From a distance of half a million kilometres Jupiter fills the sky, appearing some forty times the width of the full Moon as seen from Earth. As Galileo closes in, the giant planet grows to a huge, orange-yellow mass, its cloud belts gradually resolving themselves into a thousand swirling eddies, white hurricane spots the size of Africa, minor storms as big as Texas. As the mother ship turns on its tape recorder to catch the precious data, our traveller is now 200,000 km above the cloud-tops.

Three minutes later, the probe slams into Jupiter's upper atmosphere at about 50 km per second, just north of the equator. Of all the objects ever built by humanity, the Galileo atmospheric probe was by far and away the fastest, and its approach speed to its target now approaches that of a comet. Like a meteor, Galileo streaks across the Jovian sky, layers of its thick protective heat shield vaporizing as air friction heats the probe to an estimated 14,000°C. At this speed the artificial meteor could cover the distance from London to New York in less than two minutes. All that can be seen from the probe is an incandescent ball of plasma enveloping the probe – vaporized heat shield and hydrogen from Jupiter itself glowing in a ball of ionized atoms. Our passenger, if real, would by now have been turned into jam by the 230*g* deceleration.

Eventually, the probe slows down to something like a civilized speed. Twenty-five seconds later, less than a minute after slamming into Jupiter, the probe is falling in a parabolic arc at just 1,600 kph – less than the top speed of Concorde. No longer generating enough heat to form a plasma veil, a fantastic vista unfolds before our brave explorer as the first chute deploys – slowing the probe still further – to a stately couple of hundred kph. Seconds later a second, bigger parachute opens, and the probe is now at the mercy of Jupiter's winds.

At this altitude it is cold, and visibility is poor. Thin wispy clouds of ammonia and ammonium hydrosulphide stretch to the horizon, which is thousands of kilometres away. As the probe descends, the hazy, sunset clouds clear, revealing one of the Solar System's greatest vistas – the open skies of Jupiter, the small disc of the setting Sun perhaps peeping through the haze. By about 15 km below the entry point the pressure has climbed to a comfortable 1 bar – just perfect for an unprotected human. It is still chilly – about minus 100°C – but the temperature climbs slowly as the probe descends.

Looking up, our astronaut sees a sky of the deepest, darkest blue. Jupiter's atmosphere scatters the sunlight, filtering out the reds and the yellows. Exactly the same process operates on Earth. The hydrogen–helium air of Jupiter is astonishingly clear. Below the ammonium smog layer, you can see for hundreds, even thousands of kilometres. In the distance our astronaut can see towering cumulus clouds, tens of kilometres high, and sheets of lightning arcing between them. Within and under these clouds fall snow, sleet and rain. The same processes that govern the upward movement of water-laden air on Earth may operate on Jupiter, the fierce winds and convection currents sending ice droplets up and down into the clouds, where they grow to fantastic sizes. Among the more hazardous surprises Jupiter can throw at its uninvited guests are hailstones the size of cars.

By 80 km below the entry point the temperature has climbed to about 40°C, and the pressure to about 8 bars. Strangely enough, Jupiter, on the face of it the most alien and inhospitable of worlds, may be the only place in the Solar System where an astronaut could survive without the heavy-duty protection afforded by a full spacesuit – at least in the outer film of its atmosphere. The thick, soupy atmosphere protects against both solar radiation and the fierce ionizing storm generated by Jupiter itself. Between 60–100 km down, pressures and temperatures are in the same ballpark as those found on Earth. The "air" – a mixture of hydrogen, helium, water vapour and trace amounts of methane, ammonia and nitrogen – will not sustain life, but is not toxic either. Perhaps one day, astronauts *will* descend into this bizarre world, borne through the atmosphere by balloons, drifting through the cloudscape like modern-day Argonauts.

Taking off his helmet, encumbered only by an oxygen tank, our visitor takes a long, last look at this strange world. In some ways it appears surprisingly familiar – blue skies above, fluffy white clouds floating past. Lightning, rain and sleet in the distance. The Sun, a fifth of its familiar size, may just be glimpsed through the ammonia layers above. Could life have evolved in this strange place?

In the 1970s Carl Sagan postulated that the outer layers of Jupiter and Saturn may be home to fantastic lifeforms. Rich in hydrocarbons and supplied with abundant water, warmth and light, as well as the vital electrochemical spark of lightning, the atmospheres of the gas giants could be home to millions of floating creatures, an airborne ecology consisting of giant, jellyfish-like creatures kilometres across. Protein-like waxes, as flimsy as gossamer, could form the body structures of these giant Jovians and Saturnians. In his science-fiction novel (and subsequent movie) *2010: Odyssey Two*, the follow-up to *2001: A Space*

Odyssey, Arthur C. Clarke imagined fierce battles between aircraft-sized hunters and city-sized grazers.

However, the data sent back from the Galileo probe killed off the life-in-Jupiter hypothesis. The neutral mass spectrometer, which measured the atmosphere's chemical composition, detected "little evidence" for organic molecules around the probe. Some organics were detected, but they were simple, short-chain hydrocarbon molecules like methane, propane and butane – camping gas. No long-chain waxes, no proteins, and certainly nothing that could possibly be interpreted as amino acids or DNA. Furthermore, the fierce vertical mixing of Jupiter's layers by gigantic convection currents driven by heat generated deep inside the planet would cause anything more complex to be quickly incinerated hours after being formed, as it was dragged down into the depths. It seems as though Sagan's Jovians, like Percival Lowell's Martians, are confined to the pages of science fiction.

At a depth of 100 km we are approaching the limits of survivability for a human being. The probe's progress is now far from stately. Winds of 500 kph buffet the little spacecraft, dragging it hundreds of kilometres by its thin parachute. As it descends, temperatures climb higher and higher. Light levels fall – the blue skies turn a gloomy yellow grey, and even the titanic lightning flashes are no longer visible. Our visitor is subjected to the kind of pressures experienced by deep-sea divers, and temperatures higher than a sauna. Soon the water clouds disappear from view. Below is just the dark brown and red hellish interior. The probe starts to malfunction as the high temperatures cook its delicate instruments. Fifty-seven minutes after entering the king of the planets, around 200 km below the visible cloud-tops, the probe signs off its last data and dies as the temperature approaches 150°C and the pressures touches 20 bars.

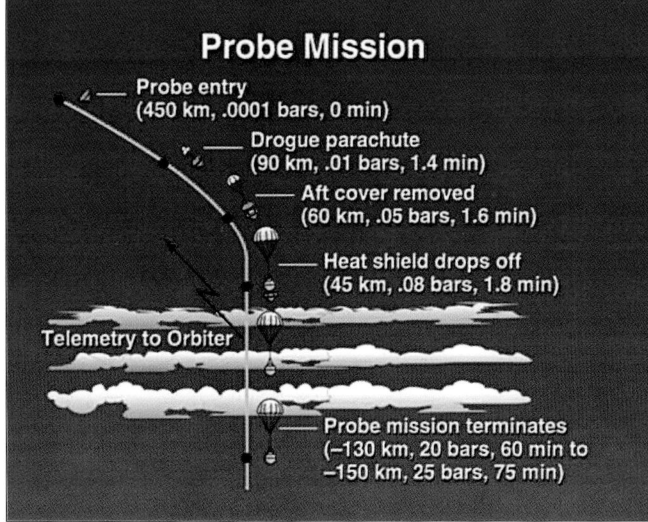

The descent path of the Galileo atmospheric probe.

Three hundred kilometres down, and the parachutes, made of hi-tech plastic, start to melt. Temperatures and pressures are now on a par with those at the surface of Venus, over 500°C and about 90 bars. Light levels are now so low that it would be virtually impossible to see anything. As the parachute disintegrates the probe accelerates rapidly, its streamlined shape easing its passage through the dense atmosphere. At 1,000 km below the probe's point of entry, the thick hydrogen air around it glows red. The temperature is now over 1,500°C, and the aluminium of the probe's body starts to melt, then evaporate. Anyone hoping to visit this place will need a thick shell of titanium for protection. Still our imaginary astronaut descends. When pressures reach the 5,000-bar level – 6 tonnes per square centimetre – the temperature is hot enough to melt even titanium. The Galileo probe stopped transmitting data about eight hours ago when its electronics fried. Now the last parts of its body start to evaporate. Sixteen hundred kilometres into Jupiter – just one-fortieth of the distance to the centre – and the Galileo probe has become at one with the giant planet, enriching its atmosphere with a few hundred kilograms of exotic metals and other elements.

Exploration below this depth will never be possible using the sort of technology we can imagine today. But we can probe to even deeper parts of Jupiter using our knowledge of the laws of physics and the observations we have from Galileo and from ground-based observatories. At depth, the hydrogen in Jupiter is compressed into a bizarre state similar in many ways to a metal. This "metallic hydrogen" conducts electricity, and a vast, deep swirling ocean of the stuff extending from maybe 16,000 km beneath the cloud-tops for a further 50,000 km is responsible for generating the planet's massive magnetic field. This metallic hydrogen is surrounded by a mantle of "conventional" liquid hydrogen – the gas rendered liquid not by cold but by pressures of millions of tonnes per square centimetre. Above the metallic hydrogen layer, scientists believe, the interior of Jupiter is lashed with helium rain, droplets of the liquefied gas at thousands of degrees, kept from exploding into vapour by pressures of millions of bars.

Near Jupiter's centre, temperatures and pressures are unimaginably high: 2 million bars, and tens of thousands of degrees. Jupiter almost certainly has a "small" rocky core, perhaps twice the size of Earth, but

This schematic cut-away view shows the components of Jupiter's ring system and its relation to the inner moonlets, which are the source of the dust which forms the rings.

so compressed that it is 10 times denser. Arthur C. Clarke once fancifully suggested that carbon compounds raining down from above could be compressed by the fantastic pressures and crystallized at the core to form a gigantic diamond. Sadly for us (but not for de Beers), this gem at the centre of Jupiter, if it exists, will be forever beyond our reach.

The atmospheric probe was probably the most robust exploring machine ever built. A replica is on display at the JPL museum, the main exhibition hall of which is dedicated to Galileo. The probe, a back-up of the real thing, sits to the side of the orbiter, its gold heat-shield glistening under the spotlights. Touching it, it is hard to imagine the fate of its sister craft – frozen, then crushed, then fried in the most dramatic end imaginable for any spacecraft.

Jupiter – King of the Planets

Jupiter reigns supreme among the nine planets, containing as it does two-thirds of the planetary mass of the entire Solar System. With today's human technology it is probably the only planet that could be detected orbiting our Sun from a planet in another star system (although recent improvements in "planet-finding" technology allow astronomers to detect Saturn – and even Uranus – or Neptune-sized objects in orbit around other suns). In its composition, Jupiter resembles a star. Its interior pressure may be as high as 100 bars. Jupiter's magnetic field is immense, even in proportion to the size of the planet: it is several times the size of the Sun, with an elongated "magnetotail", pushed out by the solar wind, that can stretch as far as the orbit of Saturn. The planet's electrical activity is so strong that it pours billions of watts of energy into Earth's own magnetic field every day.

Jupiter is endowed with 28 moons (at the time of writing – there may well be many more), a ring system, and an immense, complex atmosphere that bristles with lightning and swirls with huge storm systems. Observed from Earth, Jupiter's atmosphere is seen to be divided into dark belts and bright zones of cloud in which there are fast-moving jets and long-lived ovals. The dominant oval is the Great Red Spot, a gigantic storm which could easily swallow the Earth whole and has persisted for nearly three centuries. (Early fanciful theories about the GRS supposed that it could be some sort of solid feature – a gigantic mountain poking up through the clouds, or maybe even a floating mat of organic material.) The major belts and zones have persisted since observations began.

All this was known when Galileo arrived in 1995. The Voyagers and Pioneers before them had revealed much about Jupiter. The planet's approximate composition was known, and the photographs taken of its swirling clouds painted a picture of a weather system surprisingly familiar to any terrestrial meteorologist. Jupiter has cyclones and anticyclones, circular depressions and lightning, white fluffy clouds that are probably made of water. The key questions scientists were hoping to answer with Galileo mainly concerned Jupiter's weather. What was driving it – the Sun? Or were the convection currents and hurricanes instead driven by heat from deep below? How wet is Jupiter? Is the weather driven largely by evaporation and condensation, as on Earth? Are there towering thunderclouds? Does it rain on Jupiter?

The probe's findings

The first answers were delivered by the probe. As it plunged into Jupiter's clouds, the Atmosphere Structure Instrument found that the density and temperature of the upper atmosphere are significantly higher than was expected. This suggested the presence of a powerful internal furnace and an efficient way of transferring this heat upwards. The probe found that Jupiter's atmosphere is astonishingly clear. From Earth, the strongly coloured cloud belts suggest a murky, translucent sky. To the scientists' surprise, no thick dense clouds were found, just very small concentrations of cloud and haze materials along the entire descent path. Jupiter's air is crystal-clear. Visibility to the human eye would have been hundreds, if not thousands of kilometres. However, the probe did detect clouds in the distance – quite thick ones at that. It seems that it fell into an unusually clear area of Jupiter.

Perhaps the probe's greatest discovery was the true nature of Jupiter's winds, and where they come from. Initial results from the probe's Doppler Wind Experiment indicate that the gales below the clouds blow at 700 kph and are roughly independent of depth. Winds at the cloud-tops monitored by the Hubble Space Telescope are of similar strength. It now appears that winds on Jupiter are probably not powered by solar heating, or by heating due to condensation of water vapour – the two main heat sources which power winds on Earth. It now seems certain that the engine driving Jupiter's weather is the heat escaping from its interior.

Although water condensation does not drive the weather on Jupiter, it does play an important role on a smaller scale. Jupiter's skies are lit by enormous lightning flashes, spotted earlier by the Voyager probes. Lightning in an atmosphere means thunderstorms, and that is

This is a true-colour image of Jupiter's Great Red Spot as taken by Galileo's camera on 26 June, 1996.

indicative of regions of strong atmospheric updrafts and rain. The production of certain chemicals, including complex organic molecules of the kind that form the building blocks of life on Earth, can also depend on the amount of lightning activity. On Earth we are accustomed seeing to lightning striking from clouds to the ground. However, lightning discharges between clouds are by far the most common. On Jupiter, where there is no solid surface, lightning arcs across huge distances between towering water clouds almost indistinguishable from the familiar cumulus clouds of the skies of Earth.

What is Jupiter made of?

Jupiter was the largest of the blobs of gas and dust that condensed out of the solar nebula some 4.5 billion years ago, gathering more bulk and

Galileo used its camera filters to produce this false-colour image of the Great Red Spot using different hues to represent cloud altitude. This image shows the Spot to be relatively high, as are some smaller clouds to the northeast and northwest that are surprisingly like towering thunderstorms found on Earth. The deepest clouds are in the collar surrounding the Great Red Spot, and also just to the northwest of the high (bright) cloud in the northwest corner of the image.

mass by virtue of its powerful gravity. All the different ingredients that made up that primeval dust cloud were represented in the material that Jupiter accumulated. This makes it very different to the smaller planets, such as Earth and Mars, whose gravities were too weak to hold on to the lightest elements, hydrogen and helium – by far the commonest elements in the Universe. As a result, the terrestrial planets are mostly made of heavier elements – silicon, aluminium, iron and oxygen, the "slag" of the Solar System. From Jupiter's size and mass, astronomers have long known that it is almost entirely made of hydrogen and helium – star-stuff.

But Jupiter does not contain just these two elements. A planet made of a pure hydrogen/helium mix would be a bluish-white colour, and Jupiter is a rich mixture of oranges, reds and yellows. Swirling around in the early Solar System were countless trillions of tonnes of other materials – silicates, metals, oxygen compounds and even carbon-based organics. All these would have been present in the material sucked into the swirling maelstrom that was the infant Jupiter. Spectroscopic analysis from Earth showed that Jupiter contains significant amounts – in its upper cloud layers at least – of ammonia and methane. Some astronomers speculated that the reds and yellows could be accounted for by sulphur compounds. Jupiter was also thought to be rich in water. Deep in its interior, the chemistry of Jupiter must be outlandish – super-compressed, electrically conducting hydrogen, perhaps silicates and even pure carbon. At its heart is probably a small solid core somewhat bigger than the Earth.

So, did astronomers guess right about Jupiter's make-up? Is it a "frozen sun"? The probe's Neutral Mass Spectrometer (NMS) found that the atmosphere contains much less oxygen – present mainly in water vapour in Jupiter's atmosphere – than does the Sun's, implying a surprisingly dry atmosphere. On the other hand, the amount of carbon – present mainly in methane gas – and sulphur – present in hydrogen sulphide gas – are both more plentiful than in the Sun.

The Helium Abundance Detector experiment very accurately measured the ratio of hydrogen to helium in Jupiter's atmosphere, and found it to be near that in the Sun. But the proportion of other common elements – oxygen, sulphur, and so on – was very different to solar values. One clue as to why this should be came from the impact of Comet Shoemaker–Levy 9. Thanks to its bulk and position in the outer Solar System, Jupiter must have been considerably enriched by the impact of millions of comets over the aeons, each throwing a few gigatonnes of exotic chemicals into the Jovian soup. But accounting for all the differences – particularly the dryness of the part of the atmosphere tasted by the atmospheric probe – has proved to be a testing task for the Jovian meteorologists.

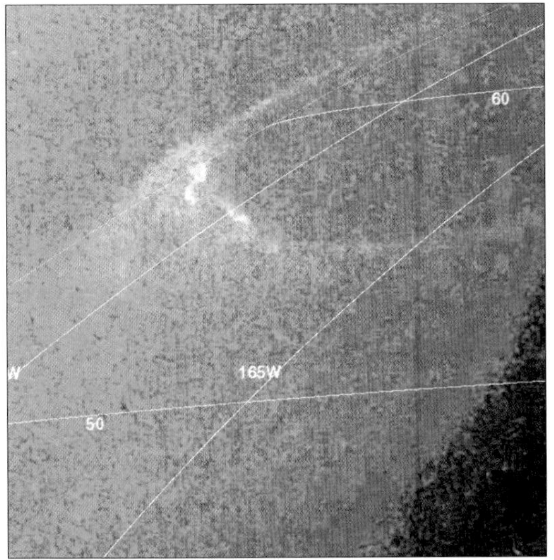

Jupiter, like the Earth, can put on a spectacular lightshow. This photograph shows the Jovian aurora, flashes of ionized gas near the planet's north pole.

Jupiter's deserts

New research has shed light on the "Jovian Sahara" discovered by Galileo. It seems that immense thermals, hot vertical winds that span great ranges in air pressure, may explain the huge surprisingly clear, dry areas near Jupiter's equator. "If you could ride in a balloon coming into one of the hotspots, you would experience a vertical drop of 100 km – more than ten times the height of Mount Everest", says Andrew Ingersoll, one of the main Jupiter atmosphere researchers working at Caltech. When Ingersoll's paper was published in *Science*, it appeared that the mystery of the Jovian desert had finally been solved. "This helps answer one of the big puzzles we ended up with after the probe entry", said Torrence Johnson at the time.

Ingersoll suggests that air moving west to east just north of Jupiter's equator is also moving dramatically up and down, every few days. Water and ammonia vapours condense into huge clouds in Jupiter's white equatorial plumes as the gases rise. Then the dried-out air plummets, forming the clear patches. After crossing those hotspots, the air rises again and returns to its normal cloudy state. After the strangely "dry" data came in from the probe scientists quickly realized that the

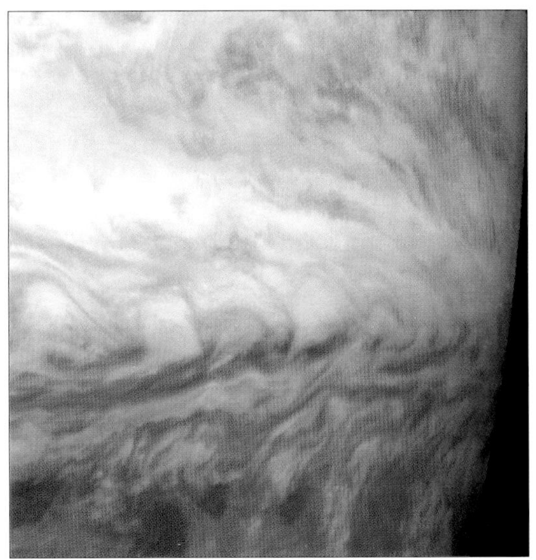

This false-colour mosaic shows a belt-zone boundary near Jupiter's equator. The images that make up the four sections of this mosaic were taken within a few minutes of each other.

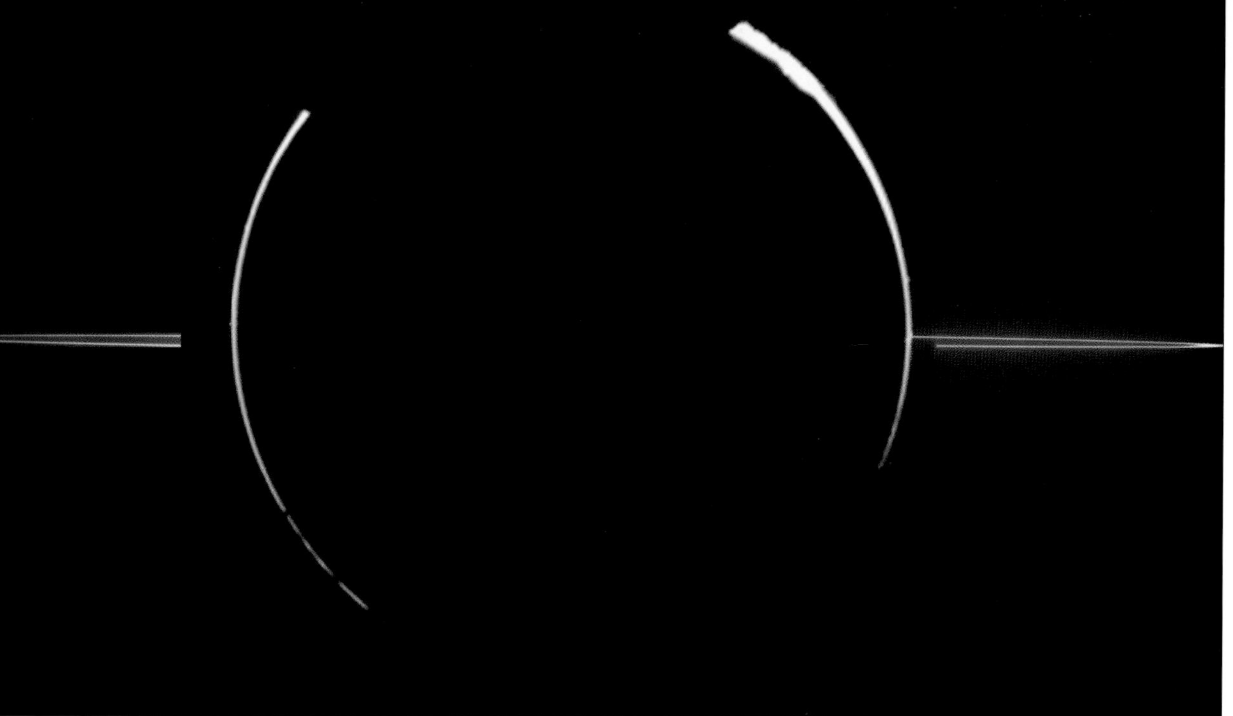

Saturn is not the only ringed planet. In fact, all the gas giants, including Jupiter, possess faint ring systems. Galileo took this picture of Jupiter's rings when the Sun was behind the planet, and the spacecraft was in Jupiter's shadow peering back toward the Sun.

point of entry was a special place. On a planet mostly wrapped in high clouds, the probe hit the southern rim of a clear spot where infrared radiant energy from the planet's interior shines through. In some ways, the dry areas where wrung-out air masses are descending resemble subtropical deserts on Earth, like the Sahara and the deserts of the American Southwest, Ingersoll said. But, unlike Earth, Jupiter has no firm surface to stop the air's fall. All the hotspots combined make up less than 1 per cent of Jupiter's total surface area, but understanding how they remain stable is important for understanding the whole planet's atmospheric dynamics.

Jupiter from space

Galileo's observations of Jupiter did not stop with the demise of the atmospheric probe. The orbiter has spent five years studying Jupiter from distances ranging from a few tens of thousands of kilometres to millions, and has answered many of the riddles the probe data left unsolved. The cloud pictures taken by the Voyager spacecraft provided some of the most beautiful, and surreal images in astronomy. The vast swirling storms of Jupiter underlie just how alien this world is; nothing like them seems to exist on Earth. Yet closer analysis by the Galileo orbiter, which has had an unrivalled view of the meteorological

Jupiter's most famous storm is the three-century old Great Red Spot. But dozens of lesser hurricanes swirl across its surface. These three classic white ovals which formed in the 1930s have occupied the band from 31–35° south ever since. In February 1998, the storms merged, forming a giant new weather system. The lower panel shows the merged oval.

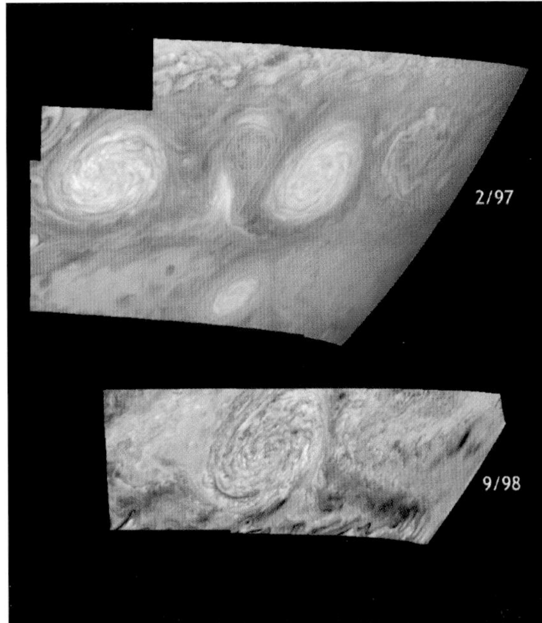

spectacle of the Solar System, has revealed that while there might be important differences between the weather of Jupiter and of Earth, there are important similarities as well.

Detailed analysis of the Jovian cloud pictures taken by Galileo, published in February 2000, showed that this remarkable planet has some quite extra-ordinary weather. Anvil-shaped clouds tower more than 50 km high, casting a pall over a hazy sky. Amid the gathering gloom, 160 kph winds whip clouds across the sky, and massive bolts of lightning arc between the clouds. Think of the worst storm ever seen on Earth and multiply it tenfold, maybe more. This is just another rainy day on Jupiter.

Jupiter experiences depressions, storms, and thunder and lightning, just as Earth does, but the heat source that drives these events is completely different. Generally, thunderstorms on Earth are small pockets in the atmosphere known as cells, containing cumulonimbus cloud, and caused by summertime heat from the Sun. Where it is warm near the Earth's surface in the summer and cooler at altitude, condensation rises and forms many cells of intense thunderclouds over a vast area. These summertime giants can persist for hours, even

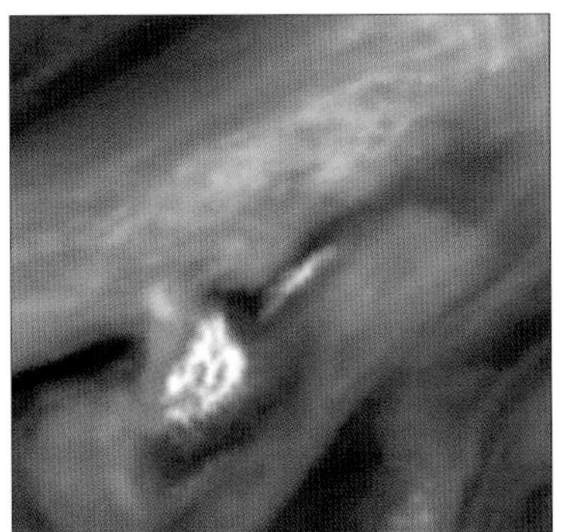

This picture of an anvil-shaped thunderstorm 10,000 km northwest of the Great Red Spot was taken by Galileo on 26 June 1996. The white cloud in the centre is a tall, thick cumulus-type water cloud 1,000 km across, and some 70 km high. If this cloud were on Earth, it would stick out into space!

days, and dump unusually large amounts of rain. On Jupiter they can last from about 12 hours to several Earth days, and produce huge quantities of rain.

Jupiter, on the other had, has no warm oceans. It is heated by the Sun – but because Jupiter is nearly five times as far from its warm glow than is the Earth, the heating effect is much, much lower: a given amount of surface area receives about one-twenty-fifth of the solar heating as on Earth. However, Jupiter has another source of heat – its vast interior reservoir of hydrogen, which is kept hot by slow gravitational collapse. Because of this internal heat reservoir, Jupiter emits nearly 70 per cent more heat than it absorbs from the Sun – a useful amount that is more than enough to drive its fearsome storms. Peter Gierasch, an expert on Jupiter's weather, explains that the physical attributes of Jupiter's vast thunderstorms are actually the same as those on Earth, except that Earth's storms are triggered by the Sun's heat, while Jupiter's are driven its own internal heat source. One part of these Jovian storm systems that dwarfs anything on Earth, says Gierasch, are the lightning bolts, which can be hundreds of kilometres long. "I wouldn't want to fly through a storm like that", he said.

A wandering planet?

One of Galileo's most startling discoveries has been that Jupiter may be a long way from home, an interloper from the far outer reaches of the Solar System and only wandering inward to its present orbit relatively recently. It has always been assumed that the planets coalesced out of the solar nebula more or less in their present orbits. But analysis of data collected by Galileo has forced scientists to rethink how – and where – Jupiter formed.

The probe found that Jupiter contains two to three times as much of the noble gases argon, krypton and xenon than one would expect had the planet formed solely from the leftovers of the formation of the Sun. It also had about three times more nitrogen than is predicted by the prevailing models of how our Solar System formed. Where Jupiter now orbits – at about five times the distance from the Earth to the Sun – is much too warm for it to have accumulated those gases in the quantities detected by the probe. Most icy planetesimals, a class of

This computer-generated 3D image is based on Galileo data and shows an equatorial "hotspot" similar to the site where the probe entered Jupiter's atmosphere. These features are holes in the bright, reflective, equatorial cloud layer where heat from Jupiter's deep atmosphere can pass through. The bright clouds to the right of the hotspot as well as the other bright features may be examples of the upwelling of moist air and condensation.

objects which includes comets, are thought to have formed somewhere between the orbits of Uranus and Neptune, 20 to 30 times the Earth's distance from the Sun. Even at that distance, however, the initial temperature of these icy bodies would have been far too warm for them to trap the heavy noble gases and nitrogen in an icy form.

It is possible that Jupiter formed much farther out, and later migrated inwards. These findings, coupled with the recent discovery of planets in orbit around other stars that appear to be much larger than Jupiter and much closer to their suns, may lend support to the idea that gas giants can migrate from one spot to another in their solar systems.

Solar systems can certainly be just about any size or shape. Until 1995 the existence of planets orbiting other stars was circumstantial at best. Since then, about sixty so-called extrasolar planets have been found (and no doubt by the time you read this the figure will be closer to a hundred). Most of these planets are giants, some several times the size of Jupiter. This is to be expected. None of the extrasolar planets has been seen directly– their presence has to be inferred from the slight wobble their gravity induces in their parent star. The bigger the planet, the bigger the wobble, and hence the more likely they are to be detected. But planet-finding techniques are being improved all the time, and recent finds have included solar systems of two or three planets. Some of these solar systems are very odd places indeed. One star, Tau Boötis, is orbited by a world nearly four times the mass of Jupiter at a distance of just 6 million km – almost scraping its star's surface. This planet – a gas giant– must be almost unimaginably strange and quite unlike anything in our neighbourhood. Heated to 1,100°C by the star's glare, clouds of liquid silicate rock would scud across dark blue skies of vaporized sodium. With that much energy being pumped into the planet's atmosphere, the storms are likely to dwarf anything seen on Jupiter. Several such "roasters" have been discovered – gas giants revolving around their stars at what seem to be infeasibly close distances. Nobody yet knows whether such systems are the norm. If they are, then the prospects for a Galaxy full of life are diminished. There is no way such planets could form so close to their suns – they must have migrated inwards from the outer part of the stellar nebula after formation. Gas giants swirling through solar systems are not conducive to well-ordered celestial mechanics. Like a bull in a china shop, lesser planets would be scattered into crazy orbits or even thrown out of the system altogether. It is possible that our Solar System, with a handful of well-behaved gas giants orbiting at a safe distance from the Sun, and the largest just in the right place to mop up much of the cometary debris heading inwards, may be an aberration to which we owe our very existence.

Jupiter's family

Jupiter is a strange and exciting world. Next to it the Jovian moons look like insignificant specks. But, as Galileo has discovered, Jupiter's family contains some very strange characters indeed.

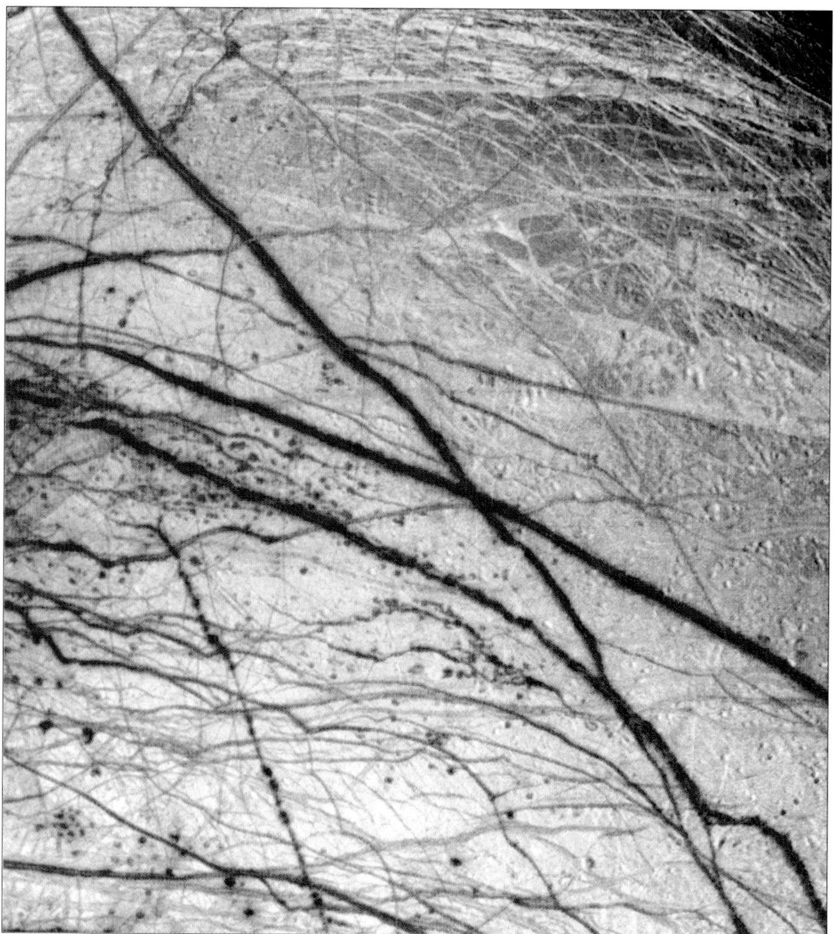

This trademark photograph of Europa is in false colour, to accentuate this extraordinary world's surface features. The picture is a composite of three images of the Minos Linea region on Europa taken on 28 June 1996. Triple bands, lineae and mottled terrains appear in brown and red, indicating the presence of chemicals in the ice. The area is about 1,260 km across.

Chapter seven

WATERWORLD

WHAT MAKES EARTH SUCH a perfect home for the warm, slithery chemistry of life is liquid water. It is possible to imagine alien biochemistries using other solvents: liquefied ammonia might do at a push, as might hydrocarbons such as butane or methane. Science-fiction writers have always had great fun envisioning alien life chemistries – silicon lifeforms, or maybe creatures based on ammonia. But despite all this, most biologists are agreed: if you really want to have a good chance of finding life, you must first look for water.

Why is water so vital? First, it is an excellent solvent. Just about everything dissolves in water to some degree, from metals (in ionic form) to substances like salts, acids and alkalis. And second, water is a liquid at temperatures that permit the existence of massive, complex molecules such as DNA and proteins that are the basis for terrestrial biota. Most scientists agree that if you could find a world with abundant liquid H_2O on the surface, or even underground, then the search for life would have a very obvious starting point.

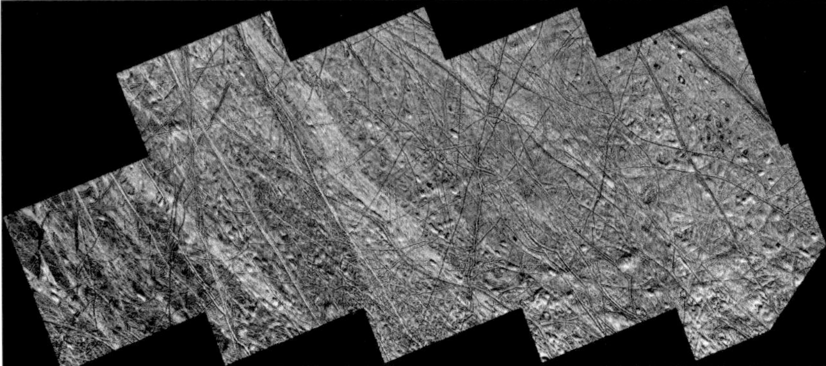

This area in the northern hemisphere of Europa displays many of the features that are typical on the satellite's icy surface. Brown, linear (double) ridges extend prominently across the scene. They could be frozen remnants of cryovolcanic activity, which occurred when water or partly molten ice errupted on the Europan surface, freezing almost instantly in the extremely low temperatures so far from our sun. Dark spots, several km in diameter, are distributed over the surface. A geologically older, smoother surface made of pure ice and bluish in tone underlies the ridge system.

But liquid water seems to be rare in our Solar System. Earth has plenty. Mars has some – mostly locked up in the polar caps as ice. As for water vapour, Mars has a few wispy clouds of ice crystals, but it has been calculated that if every molecule of water in the Martian atmosphere were to simultaneously and instantly condense out as liquid, the resulting deluge would amount to just a quarter of a millimetre of rain – not enough even to moisten your skin. (However, in the summer of 2000 new results from the orbiting Mars Global Surveyor spacecraft seem to suggest that channels on the Martian surface have been carved by flowing water in the relatively recent past.)

What about the outer planets and their moons? Scientists have known since the 1950s that a huge amount of water ice was present on Europa. Near-infrared photometry provided the first clue to the surface composition of Europa long before the first spacecraft were dispatched to the outer Solar System. Two scientists, Gerard Kuiper 1957 and Vasilis Moroz in 1961, concluded, using this technique, that both Europa and Ganymede had water-ice surfaces.

At first glance Europa looks unspectacular. With just under a hundredth the mass of the Earth and a diameter of 3,138 km, Europa is almost exactly the same size as our Moon, but rather lighter. The twin Voyager probes, despatched to Jupiter and beyond in the 1970s, gave us our first ever close-up view of Europa. The Voyager pictures showed pale yellow icy plains with red and brown mottled regions. Some areas are pristine white, unmarked by craters and uncoloured by dust of any kind. Careful measurements showed that no feature on the surface differed in altitude from any other by more than a few hundred metres. Europa is the smoothest object in the Solar System.

But the strangest thing about Europa was its lack of craters. Every large solid object in the Solar System has been battered by flying space debris for billions of years. The entire surface of our Moon is an interlocking mosaic of craters, few of them less than three billion years old. The fact that so many of the lunar craters are still visible tells us that the surface we see today is ancient – unworked by the forces of erosion by the rivers and winds that constantly scour and reshape the Earth's surface (a process that happens to a lesser extent on Mars). It also indicates a world that is geologically dead, without the active volcanism and mountain building that are constantly reshaping the Earth's surface. Scientists use the number of craters as a rough marker of the age of a planet's surface. This method shows that Europa's surface is very young indeed. Less than a dozen impact craters were identified from the Voyager pictures – and remember, this is a world with no wind or rain to scour the surface clean of pockmarks every few million years, as on Earth.

Hal Levinson of the Southwestern Research Institute in San Antonio, Texas, has used data and computer models of the dynamics of the evolution of cometary orbits to conclude that on average a 20-km impact crater should form on Europa's surface every million years. We can see about a dozen such craters on Europa, so, it can be concluded, the surface is about 12 million years old, and much younger in the crater-free regions. So, Europa's surface appears to be – compared with the age of the Solar System – very, very young, and this has tremendous implications. How could Europa remodel itself in a matter of a few million years in a place where there is no erosion, erasing the bombardment history that has scoured the surface of almost all the airless worlds in the Solar System?

Back in the Voyager days, some people thought they knew why. The top of Europa is made of ice, they said. Ice is frozen water. Warm that ice up and you get a liquid, this liquid could be the key to Europa's resculpted surface. Now, how could a little world three quarters of a billion kilometres from the Sun get warm enough to reach a sweltering 0°C? Tidal heating and volcanism might be the answer.

Of all the planets in the Solar System, the only one known in the 1970s to be geologically active was Earth. On our world the crust is constantly reworked in a series of uplifts and collisions driven by a reservoir of radioactive heat keeping things warm underneath the crust. The "unifying theory" that explains everything from the distribution of earthquakes and volcanoes to the topography of the ocean floors and mountain ranges on our planet is called plate tectonics. This theory, first postulated in the 1960s, states that the surface of the Earth can be regarded as a jigsaw of about a dozen rigid "plates" up to 100 km thick which move relative to one another, propelled by convection currents in the mantle underneath.

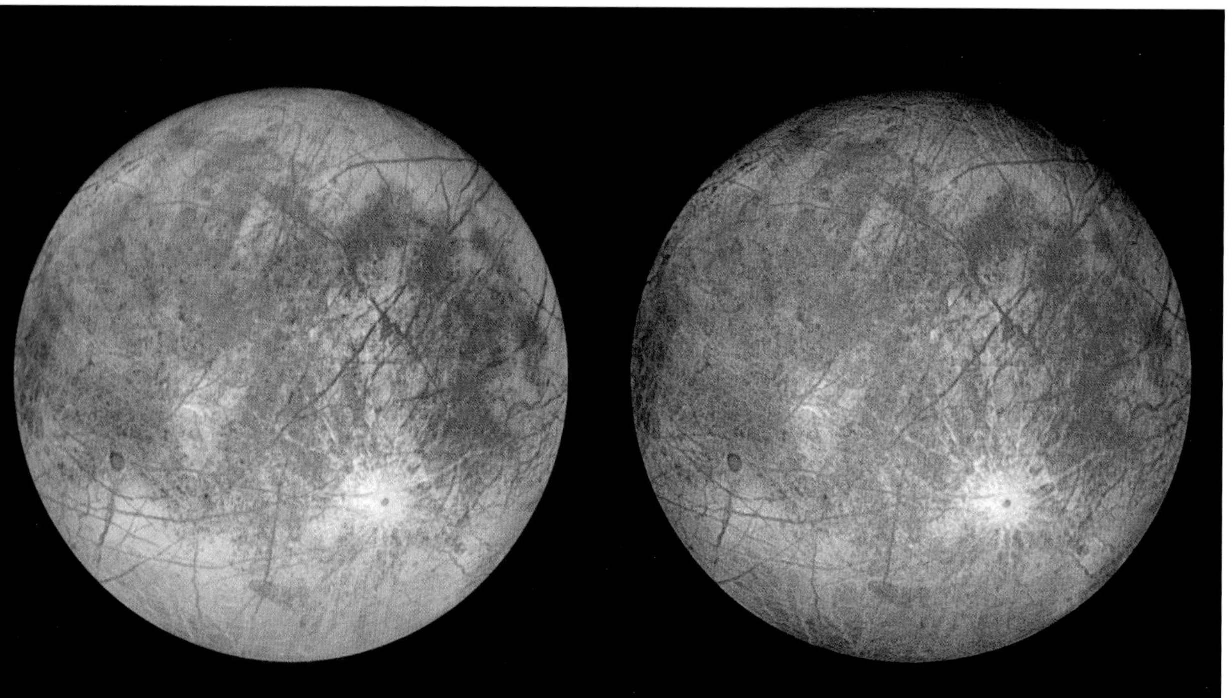

This image shows two views of the trailing hemisphere of Europa. The left image shows the approximate natural appearance of the moon. The image on the right is a false-colour composite version combining violet, green and infrared images to enhance colour differences in the predominantly water-ice crust.

The theory explains major landforms such as mountain ranges, volcanoes and undersea trenches as features of plate margins. For instance, when plates collide, material is pushed upwards, forming mountains. The relative movement causes stresses, which generate earthquakes. When plates pull apart, molten rock wells up from below to fill the gap; this can be seen along ocean ridges such as the one that runs north–south down the Atlantic Ocean. However, it seems that there is a critical size for plate tectonics to occur: if a planetary body is too small, like Mars or the Moon, the internal heat soon leaks away and any volcanism ceases. It looks as though the Moon's crust was never broken into plates by convection currents below. On Mars, the great volcanoes of the Tharsis plateau, including Olympus Mons, the biggest volcano in the Solar System, indicate a turbulent geological past. But Mars too appears to have been "frozen", its crust never having split up into active and mobile plates, as happened on Earth.

But there is more than one way to make a volcano. Even if a planetary body is too small to generate much heat internally, it is possible that tidal forces could do the trick. And Europa is subjected to almighty tidal forces. It orbits the hefty Jupiter in just over three and a half Earth days, at a distance of 671,000 km. Jupiter's high gravitational pull is not enough in itself to generate a tide – with one face locked towards the parent planet, Europa could simply be pulled into a permanent rugby-ball shape, with no consequent flexing or heating. But calculations performed in the 1970s by Stan Peel and other astronomers showed that the complex gravitational dynamics of the Jovian system, and particularly the gravitational interactions of Jupiter and the four main satellites, should raise tides on Io and Europa, at least, that are high enough to generate a lot of heat. The case for a warm Europa was strengthened when the Voyager probes saw the erupting volcanoes on Io – Jupiter's innermost moon. A planetary body that small should be geologically totally inert; the fact that Io breathes fire is proof that tidal volcanism in the inner Jovian System is a reality.

In fact, Europa's "tides" must be many times more powerful than anything seen on Earth, and these tides must be rising and falling in ice and solid rock. Like a piece of metal bent back and forth, the silicate rocks and the ice crust above are warmed by this gravitational squeezing, just as the metal in a paper clip gets warm when you bend it repeatedly. Under the ice, the rock may be hot enough for volcanoes to form, as on Europa's neighbour, Io. These volcanoes would be enough to melt the ice locally. But could the ice be molten under the entire surface of Europa? Could this world, so far from the warmth of the Sun, be home to an immense, planet-wide ocean? This would certainly help to explain its puzzling topography, but as yet there was no firm evidence – just plenty of speculation.

"We knew about the ice from ground-based observations", says Brown University geologist Bob Pappalardo, one of the main Europa investigators. "We've known that since the sixties, and people were talking about oceans under the ice as far back as 1971." Galileo chief scientist Torrence Johnson adds, "For Europa, we had to guess at how thick the ice crust was, so we arbitrarily decided that it was probably fairly thin. Its overall density is three grams per cubic centimetre, and so we said, 'Well, if it's got an ice crust greater than 75 km thick then it would have an ocean underneath.'" So, before Galileo, the evidence for an ocean was compelling but circumstantial. Volcanism had already been confirmed on the innermost of

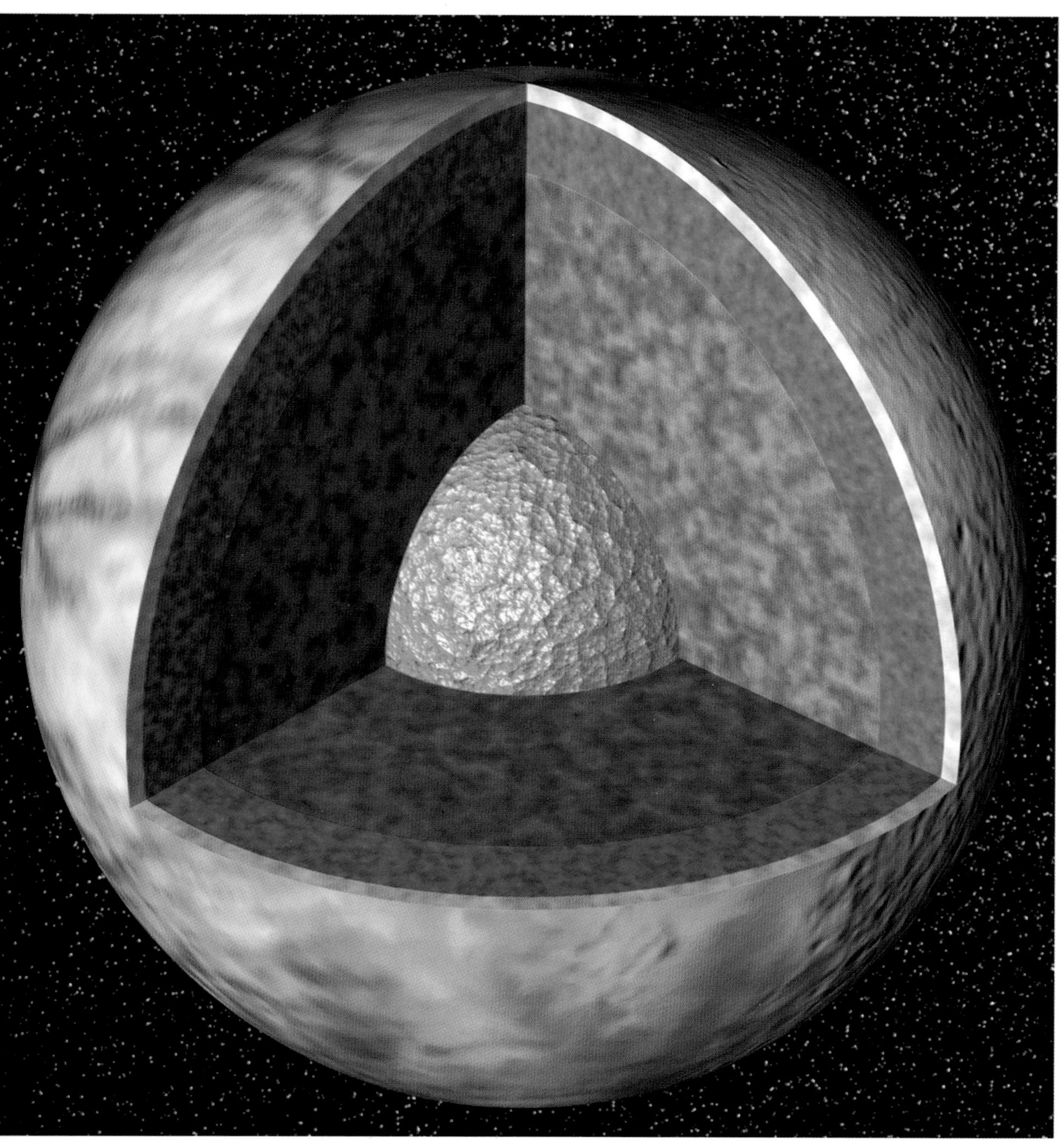

This model of the interior of Europa is the current "best guess" of how this moon is constructed. An outer ice layer about 5–20 km thick encloses a water ocean up to 100 km deep. This in turn covers a silicate mantle, which encloses a metal-rich core. We do not know if tidal heating is strong enough to melt the silicate mantle and produce volcanism – as on Io. But all the data from Galileo suggest strongly that Europa does have a liquid water ocean – only the second such ocean in the Solar System to be discovered.

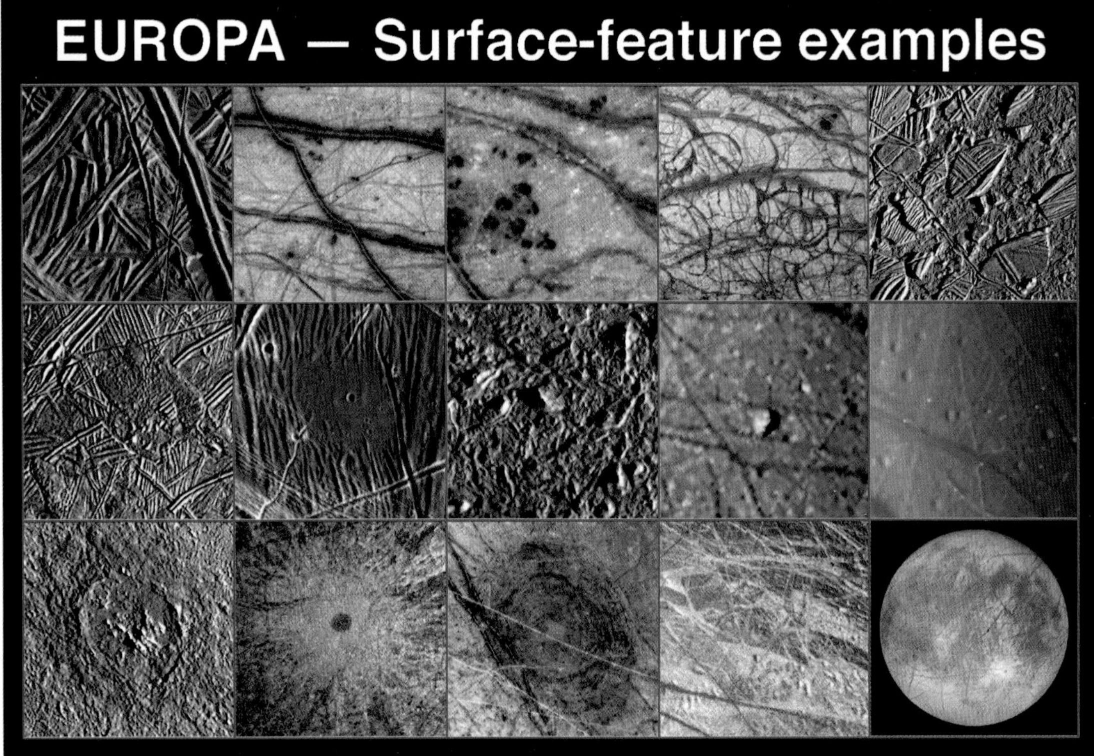

These 15 frames show the great variety of surface features on Europa, which have been revealed by Galileo during its first six orbits around Jupiter from June 1996 to February 1997. North is to the top of each of the images.

Jupiter's "big four", Io. Io was far too small to generate this much heat on its own; its volcanism must be produced entirely by the tidal kneading it was getting from mighty Jupiter. Europa was only a little farther out; maybe it was being heated as well.

The possibility of a Europan ocean was one of the most exciting astronomical mootings for decades, although it was a long way from being proved. Arthur C. Clarke wove the concept into *2010: Odyssey Two*, in which Europa is not only home to a sub-surface ocean, but also to life. Any undersea volcanoes, he reasoned, will generate vast quantities of dissolved and particulate nutrients – just as the abyssal volcanoes on Earth's ocean floors churn out compounds rich in sulphur and metals. Such thermal and chemical energy could easily fuel an entire ecosystem, protected from the icy cold of space (and from the lethal radiation of the near-Jovian environment) by a thick cover of ice. Clarke envisaged a

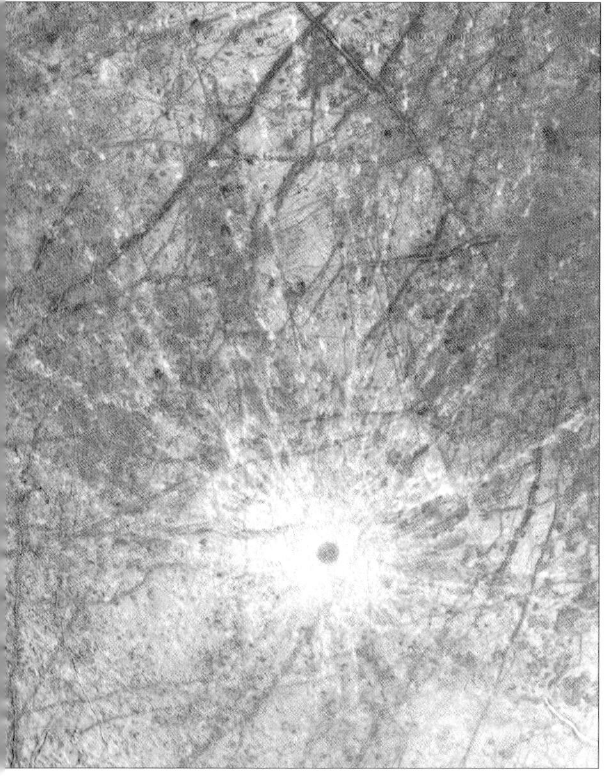

This enhanced-colour image shows the young impact crater Pwyll, just below the centre of the image. The diameter of the central dark spot, ejecta blasted from beneath Europa's surface, is approximately 40 km, and bright white rays extend for over 1,000 km in all directions from the impact site. These rays cross over many different terrain types, indicating that they are younger than anything they cross.

whole food chain, from microbes and plankton to Europan analogues of fish and seaweeds, and even semi-intelligent dolphin-like creatures that used biophosphorescence to see. Clarke's vision was a fantasy – founded on what little scientific evidence was then available – but it was a bold guess. Thirty years ago you would have got very long odds indeed on there being life anywhere in the Jovian system, let alone on frozen Europa. Many scientists now suspect that there might be something at least microbial lurking under Europa's surface, although the jury is still very much out on fish – let alone dolphins. The picture of a Europa teeming with animals and plants was a nice idea – life in a thermos flask, a baked Alaska in reverse. But as Galileo neared its target, no one had any idea whether it was true.

That didn't stop the excited speculation. If there is one thing that keeps people interested in space, it is the possibility of finding extraterrestrial life. Pond slime, microbes, anything which shouts "You are not alone!" would not only be the story of the century, it would also push astronomy and space exploration back to the top of the agenda and guarantee a steady flow of government funding. Find even a hint of an ocean on Europa, and the chances were that Congress would give the nod and a wink to a follow-up mission to orbit the little moon, and maybe even land there.

Voyager's photographs of Europa were tantalizing. What would Galileo see when it made its first, long-distance pass of Europa in June 1996? Arizona State geologist Ron Greeley said that NASA was desperate to get a peek at Europa, and get a close-up view of the strange structures glimpsed by Voyager all those years before. He stated at the time, "A major goal of Galileo's studies of Europa is to search for signs of past or current activity to help us answer the question, 'Is there a liquid ocean?' We want to go back to some of the areas which suggest soft ice or liquid water under the ice, and test some of the questions we are asking now."

How easy would it be to "see" an ocean under the ice? Galileo is equipped with some powerful instruments, but unless the probe was very lucky indeed, finding conclusive and direct proof of a sub-surface water ocean would be hard. The scientists reasoned that there was an outside chance of Galileo flying over an erupting geyser. This wouldn't prove that there was an ocean under the ice, but it would show that Europa is geologically active and had enough heat to melt at least some of the water some of the time. Better still would be to glimpse patches of open water. If the ice surface was very active, it should resemble the Arctic pack ice on Earth, with huge ice rafts separated by ridges where the floes collide, and perhaps linear tears where they are torn apart. Voyager had seen plenty of linear structures on Europa – long, straight, dark bands that look like volcanic dykes – ridges of solidified lava that form linear features on the surface. Could these be places where liquid water is welling up from below, squeezing its way through cracks in the ice crust?

The first images were taken when Galileo swept past, 160,000 km from Europa. This was only 20 per cent closer than the Voyagers had approached, but Galileo's digital camera is a massive improvement on the Voyager imaging system. And the pictures were stunning. On 13 August 1996, after the painstaking tape playback routine had been performed and the pictures received, a jubilant NASA rushed out a triumphant press release, "Jupiter's Europa Harbours Possible Warm Ice or Liquid Water". NASA posted the results of the flyby, showing unprecedented close-up detail of Europa's alien surface. Even at a maximum resolution of about a kilometre and a half, Galileo's snapshots of an area around Europa's north pole were as indicative of an ocean as anyone in NASA could have hoped for. The images showed ice-flow features, and surface deposits of coloured material that suggested geyser-like eruptions. Galileo also photographed vast "triple-lane highways" crisscrossing the Europan terrain.

No wonder that when Europa joined the (very) short list of places where life might, just, be found, this moon rapidly became a major focus of attention for the Galileo mission. Torrence Johnson remembers the day the first images came in from Galileo, images which seemed to show a crazy-paving ice crust. "I remember looking at that picture in my own office and saying, 'That sure as hell looks like icebergs to me', and 'This is going to be the picture that confirms we had something melted there.' And I called several people in to take a look at it. People across the country and all over the world were doing the same thing, and then frantically e-mailing people back and forth about what they thought it might mean."

Across most of the surface the ice appears white, or mottled brown and yellow. One possible source of the yellow colouring is sulphur squirted across space by Io's volcanoes. In the photographs some areas appeared to have rocky deposits – a puzzle if there was a contiguous ocean under the ice. And other areas appeared to be clear, looking just like recently frozen water. One explanation of the triple highways is that the ice cracked along a deep fault, and then a "lava" of dirty, gravel-filled water or maybe slushy ice flooded up onto the surface where it immediately froze in the vacuum, sealing the fault. This would explain the colour variation along these structures – from blue ice, which presumably results from eruptions of pure water, to the dark bands, suggesting contamination. From the lack of craters, these features could have formed as recently as yesterday. Certainly few of them could have been more than a million years old. The surface of Europa was shaping up to be a very interesting place. Bob Pappalardo remembers: "Even back then we saw this interesting place just north of the equator that looked like an upside-down bunny, and retargeted Galileo go there and take a closer look."

NASA chief Dan Goldin was moved to comment on the findings himself. "These fantastic new images of an icy moon of Jupiter are reminiscent of the ice-covered Arctic Ocean – the lack of craters, the cracks and the signs of movement all indicate that this might be young ice on a dynamic surface." He added, "The pictures are exciting and compelling, but not conclusive." Even at 160,000 km – nearly half the distance from Earth to its Moon, Galileo's solid-state digital cameras were taking some amazing snapshots. Most prominent were the massive, linear features that criss-crossed Europa's surface like Percival Lowell's Martian canals. Some were as straight as the roads of the American Midwest, continuing, undeviating, for many hundreds of kilometres across the polar terrain. The scientists were amazed, and searched for terrestrial analogues for the alien landscape they were seeing.

"The fracture patterns – extending a distance equivalent to the width of the Western US – dwarf the San Andreas Fault in scale and length", said Ron Greeley. The longest of the freeway-like triple bands extends for several thousand kilometres – the width of North America. Europa's surface has the longest linear features seen anywhere in the Solar System, dwarfing even the great canyons of Earth and Mars. Despite their immense length, the triple bands were not so dramatic in the third dimension – they represented cliffs and troughs at the most a few hundred metres high.But with just one long-distance flyby, Europa suddenly became the focus of the entire Galileo mission. "It's a marvellous place", said Greeley.

But to prove the existence of a Europan ocean would need many more observations. The best the Galileo team could hope for was to see moving ice on the surface. Galileo is not equipped with a movie camera, but a series of panoramas taken over days, weeks and months of the same spot could, just conceivably, show changes in the surface. This would have been achievable with a functional High Gain Antenna, with its ability to send hundreds of pictures at a time – in real time – rather than relying on the clumsy combination of low-gain antenna and tape recorder. Without the HGA, no movies of Europa could be filmed and sent back to Earth.

Active cryovolcanism could also be spotted from space, the scientists hoped. In Europa's one-seventh gee, a geyser the size and power of Old Faithful in Yellowstone Park would shoot 40 km into space! Galileo was going to be swooping to within 300 km of the Europan surface during the flybys planed for 1997. And there were other measurements that Galileo could take, of less tangible phenomena. The spacecraft's magnetometer, run by Professor of Space Physics at UCLA Margaret Kivelson and her team, would be able to detect any magnetic field the little moon might possess, and tell her something about its source. In the end, it was this magnetometer, not any amount of dramatic pictures, that would answer once and for all the question of what lay in Europa's mysterious depths.

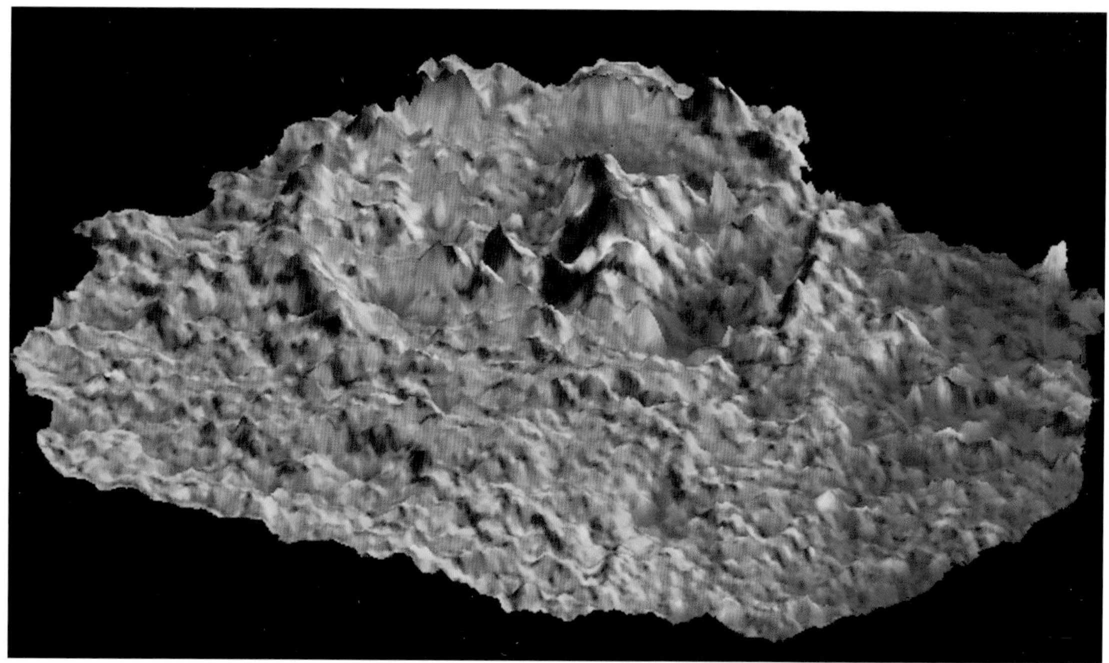

Europa's surface is young. We know this because it has very few craters. This computer-generated image shows one of the few impact structures on the Europan surface, a crater named Pwyll. The structure of the crater suggests that whatever hit Europa punched its way into a slushy or liquid layer beneath the ice.

Chapter eight

EUROPA REVISITED

Galileo's observations of Europa in 1997 were included by *Science* magazine in its top ten achievements of the year – and justifiably so, because they may make us rethink our ideas about the origins of life. Galileo's primary mission ended just two years after its arrival at Jupiter in December 1995. But because the spacecraft was still functioning perfectly, NASA decided to fund an extended mission for another two years, concentrating on Europa and Io. Because frozen water and volcanoes, respectively, dominated these two worlds the mission was dubbed "Fire and Ice". The 1997 flybys and those during the extended mission have produced some of the most spectacular results in the whole of Galileo's odyssey.

Conamara

In January 1997 Galileo beamed back to Earth a picture of an area about the size of Sri Lanka lying close to the Europan equator – the "bunny-shaped area" that Bob Pappalardo had spotted on the first, distant flyby, by now given the official name of Conamara Chaos. This area, photographed from a distance of some 60,000 km, was a city-sized patch of mottled, chaotic-looking terrain not far from Pwyll, one of the few large impact craters on Europa. Weeks later, Galileo had a chance to fly in close to this intriguing patch of ice. From a distance of just 586 km, Galileo was able to take high-

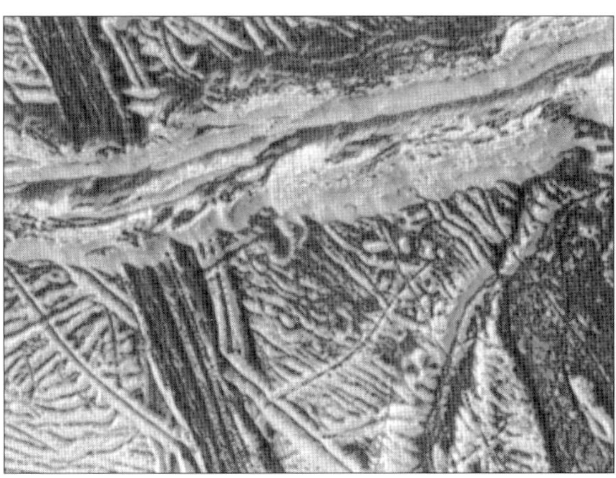

This high-resolution image of Europa, taken by Galileo in March 1998, shows a dark, relatively smooth region at the lower-right hand corner of the image which may be a place where warm ice has welled up from below. The region is approximately 30 km^2 in area. An isolated bright hill stands within it.

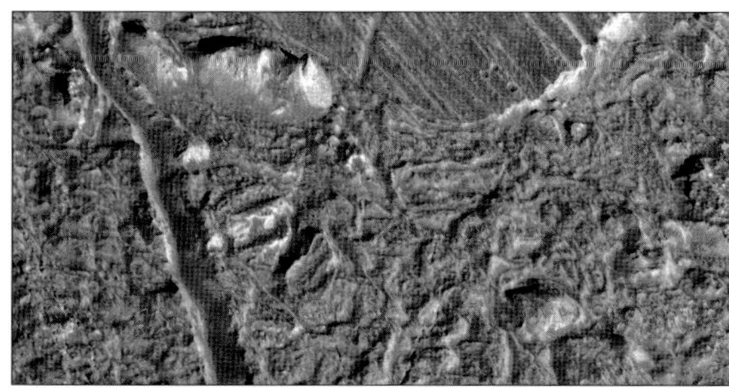

This close-up view of the Conamara Chaos region of Europa's crust was taken on 16 December 1997. The resolution is 9 m per picture element.

resolution pictures of what has now become one of the most famous landscapes in the Solar System. Conamara looks like frozen Arctic pack ice. Hundreds of blocks are assembled in a matrix, protruding some 100—200 m above the surrounding terrain. Everywhere the ice is broken into slabs, perhaps 100 m thick, which jostle against one another.

In April the JPL scientists started to analyse the results of this sixth Europa flyby. At Conamara, it looked as though a huge area has been broken up into a series of rafts which had churned around on a turbulent sea beneath, before the whole thing froze solid, and rather rapidly. Ron Greeley imagined how he would interpret the scene if he was looking at a picture of Earth. "These blocks of ice are similar to those seen on Earth's polar seas during springtime thaws", he said at

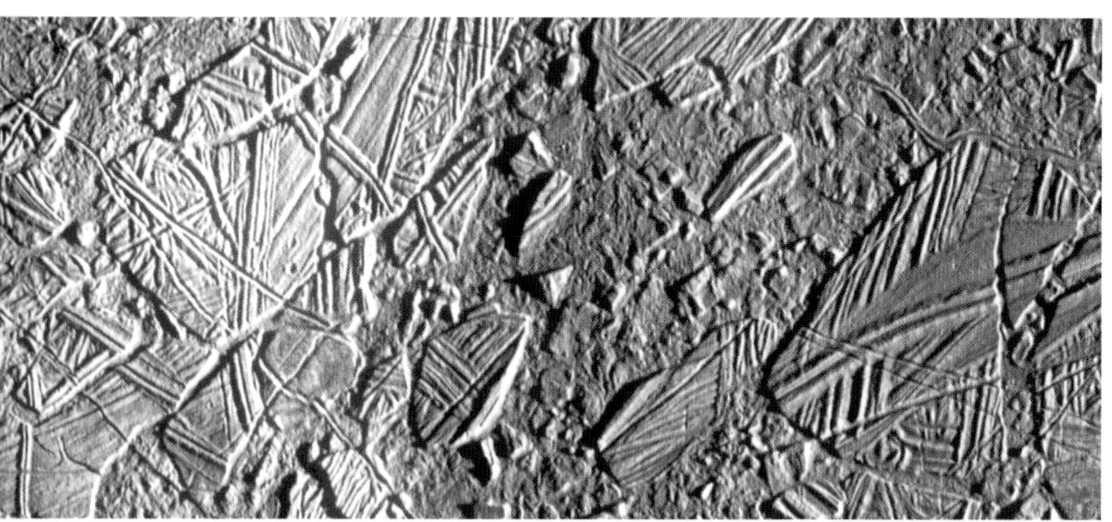

This is one of the most famous images taken by Galileo, and shows the Conamara "ice raft" region of Europa. Here, the surface has been fractured into a series of blocks, ice "rafts" apparently floating on a liquid or slushy sea underneath, which has subsequently solidified into an ice matrix.

This image was taken under "low-sun" illumination – the equivalent of taking a picture from a high altitude at sunrise or sunset. The length of the shadow cast by a feature (e.g. a ridge or hill) is indicative of that feature's height. In this image, ridges and irregularly shaped hills, ranging in size from 5 km across down to the limit of resolution (0.44 km/pixel) can be seen. Measurements from shadow lengths indicate that features in this image range from tens of metres up to approximately 100 m in height.

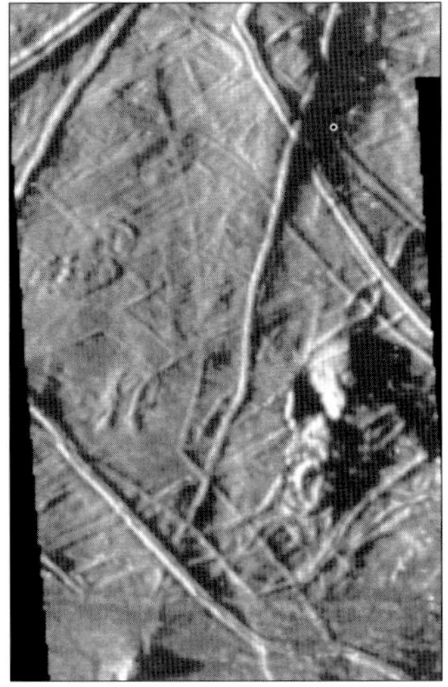

the time. "The size and geometry of these features lead us to believe that there was a thin icy layer covering water, or slushy ice, and that some motion caused these crustal plates to break up." Not only had the ice rafts rotated horizontally, they had also tilted vertically, sometimes forming cliffs many tens of metres high where they abutted a neighbouring raft. Possibly the ice was super-thin at this point, perhaps less than 300 m thick. Or maybe even thinner – some speculated that these icebergs floated on a sea that was only 30 m or so below the crust. Ron Greeley believed that the evidence for the ocean was now looking ever more conclusive.

According to planetary geologist Mike Carr of the US Geological Survey, the "icebergs" are, if not proof, then the "smoking gun" – the telltale sign that an ocean had at least once existed beneath the Europan ice. "I think the chances are pretty good that there's liquid water down there ... you could be looking at an 80-mile-deep [130 km] ocean," he said. Sceptics continued to argue that the movement in the ice was anything but sudden: there were no icebergs merrily bobbing up and down on the ocean waves. Instead, slow convection currents in solid ice can cause the kind of motion that would produce the observed surface features. Under enough pressure, "warm" ice will flow, as happens with terrestrial glaciers. Some argued that what we were seeing in the chaotic terrain was no more than the Europan equivalent of crustal folding on Earth, the ice deforming like rock (indeed, solid ice *is* rock) under tremendous pressure, driven by heat from below that is substantial yet never enough to convert the ice into liquid.

But Carr is not convinced. He thinks that only a subsurface ocean can explain what we see on the surface of Europa. "I am very, very sceptical about whether this could be explained by solid-state convection. These blocks drifted around the surface and appear to be plunging into the

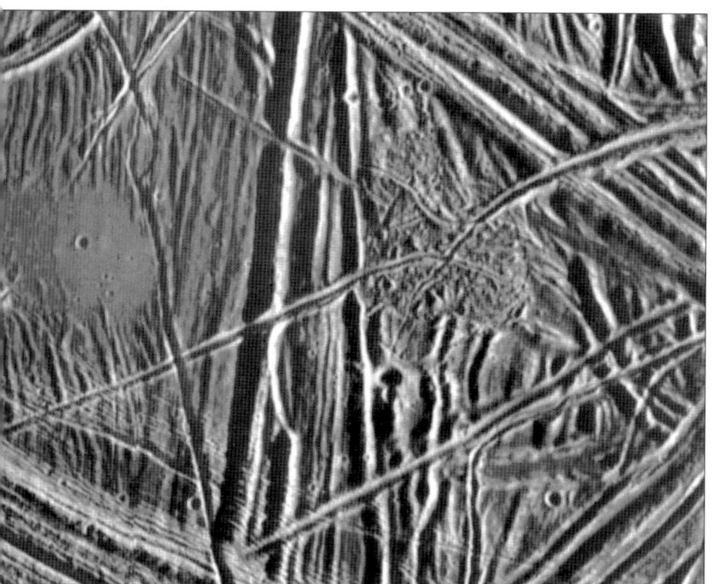

This photograph of the Conamara terrain shows a small, almost perfectly smooth area that could be made from recently frozen ice.

surface. Water appears to have crept up on top of the blocks and splashed around." Some of the most intriguing pictures taken on this flyby show what Carr and colleagues in a 1998 *Nature* paper describe as a "puddle" – a small, very dark, very smooth area. Could this be an area of very thin ice on the surface, a tiny "window" on the vast ocean below?

Bob Pappalardo advocates an ice layer "15–25 km thick – that is what the thermal models argue for." He points out that a thicker ice shell would actually generate more, not less, tidal heating, as liquid water does not heat up when it is squeezed. Nevertheless, he was as excited as anyone when he first saw the Conamara pictures; "I thought, wow! Those look like cracked plates. They look to me rather like the fretted regions on Mars, where ground has collapsed due to loss of material from beneath. I believe that 'icebergs' is a misleading term, however, as it invokes ice chunks floating around in liquid water, which may or may not be the case." Galileo also photographed structures called maculae, concentric circular structures first spotted by the Voyager probes that were a mystery to the scientists poring over the images. These structures may represent the tops of warm masses of ice brought to the surface by convection. Closer still, Galileo was able to pick out structures as small as a football field. In one image, showing a region little more than 50 km^2 in area, Galileo's cameras photographed a complex pattern of cross-cutting ridges and grooves, presumably resulting from some sort of movement. In other places, dark splodges spread across the ice. These were presumed to be "lava flows" (of liquid ice, that is), the solidified extrusions from below.

Torrence Johnson was sure that Galileo had found something spectacular. "I was pretty convinced – and I am a pretty cautious person – that we had found an ocean. There was still the question of whether or not we could have had an ocean that had frozen over, but then you have to think of Occam's razor – we have had an ocean there for 90 per cent of Europa's history, what is the chance that it froze up yesterday, just before we decided to go there? Pretty slim. It's not credible, but not impossible either. Lots of funny things happen in nature. Those icebergs convinced me that sometime in Europa's history, probably very recently, it had liquid water underneath the surface." The first results of 1997 certainly boosted the case for a Europan ocean, but they did not prove it. From closer up than ever before, Galileo had revealed a world of staggering complexity, a truly alien place quite unlike anything anyone had ever seen before.

Other features suggestive of an ocean (or at least an active subsurface layer) are the thousands of structures on Europa's surface termed "lenticulae" by JPL. These resemble blobs of solidified magma such as are seen on the Earth, but made of frozen water rather than rock. "These are strong evidence that Europa's ice shell is convecting, or at least did so in the geologically active past", Pappalardo says. In many places, what look like lava flows can be seen covering the surface. Again, this is evidence for an ocean having existed in the past, but not necessarily existing today.

Never mind icebergs floating around. What if the entire Europan surface was afloat on the water below? Is there any way that we could tell whether the surface we see is able to move independently of the rocky interior? According to the US Geological Survey's Paul Geissler, there is – and he claims to have discovered that the Europan surface is "decoupled" (that is, free to move) from the silicate/metal interior. Careful analysis of photographs taken by earthbound telescopes and by Galileo have shown that Europa's surface is *not* locked to Jupiter: instead, it turns slowly every 10,000 years or so. This movement would be imperceptible for anyone standing on the surface or orbiting in a space probe, but it has tremendous implications. This relative movement is possible only if the surface is able to flow *en masse* over the rocky surface. "We have found evidence that the crust is mechanically decoupled from the interior, just as it should be if it is underlain by an ocean", said Geissler after he published an article on the subject in *Nature* a year after the 1997 flybys.

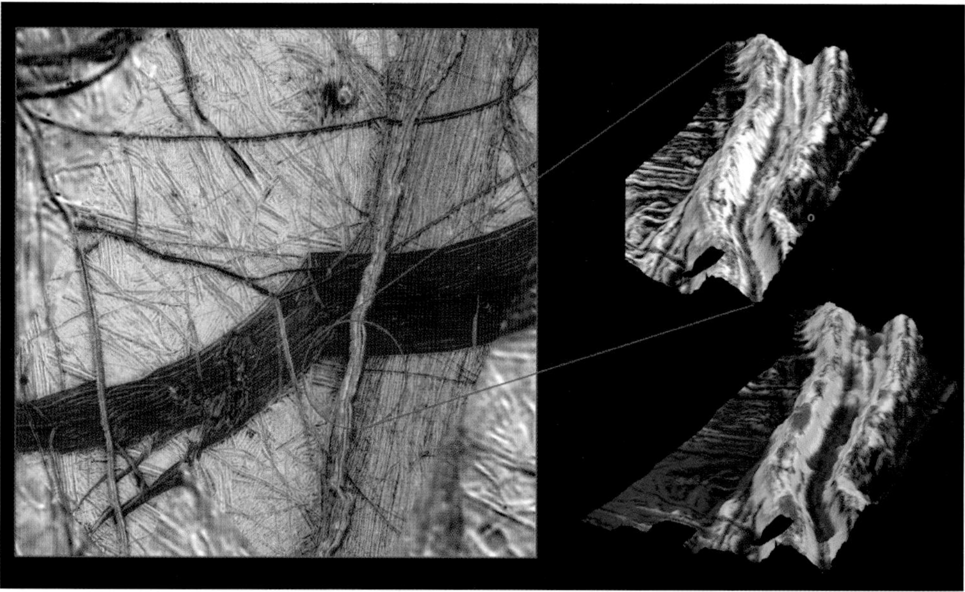

These images reveal the dramatic topography of Europa's icy crust. An east–west running double ridge with a deep intervening trough cuts across older background plains and the darker wedge-shaped band. The numerous cracks and bands of such terrain may indicate where the crust has pulled apart and allowed dark material from beneath the surface to well up and fill the cracks. A computer-generated three-dimensional model (upper right) shows that bright material, probably pure water ice, forms the ridge crests and slopes while most dark material (perhaps ice mixed with silicates or hydrated salts) is confined to lower areas such as valley floors. The model on the lower right has been colour coded to accentuate elevations. The red tones indicate that the crests of the ridge system reach elevations of more than 300 m above the surrounding furrowed plains (blue and purple tones).

Several months before, the Doppler measurements of Galileo's radio signal had shown that the moon has a crust about 160 km thick, through in itself this said nothing about whether that crust was solid, partially solid, or mostly liquid. If the overlying ice were just a few kilometres thick – the upper estimate, in the eyes of most of the project scientists – then in proportion this would be a skin thinner than that of an apple. In 1997, most astronomers thought that the Europan crust was considerably thinner than that – maybe on average less than a kilometre, though in some places much thinner, and in "cold spots" maybe several times greater. But the conclusion was clear – the Solar System had another ocean. "How often is an ocean discovered?" mused JPL scientist Richard Terrile. "The last one was the Pacific, discovered by Balboa five hundred years ago."

During the extended mission the evidence piled up for an ocean. The first of the eight extra flybys of Europa took Galileo closer in than ever before – to an altitude of just 200 km. From this distance objects the size of a pick-up truck were visible to Galileo's camera – the best resolution so far on the mission. Conamara Chaos was revisited. Skating above the surface, Galileo's camera revealed that the matrix in which the ice rafts were embedded has a rich, chunky texture, implying that when this region was formed open waters had swirled with lumps of ice ranging in size from a metre or so to several kilometres. The buoyant rafts – the icebergs – ride high in this matrix. Close up, this jumble of granite-hard ice looks like a warehouse roof.

Heat sources

The main source of the heat needed to keep a substantial layer of ice liquid in Europa would be tides generated by Europa's interaction with Jupiter, Io and Ganymede. Radioactive heating from the interior could also contribute to the melting. The tides are strong enough to cause substantial deformations in the icy crust of the moon. If the crust is solid, then the tides should amount to no more than a metre or so. If there is a thick liquid layer, however, then the surface of Europa could rise and fall by several tens of metres, maybe more. So how big are the Europan tides? The instruments aboard Galileo are not sensitive enough to measure any tidal rising and falling of the ice crust. A liquid ocean would mean that the outer

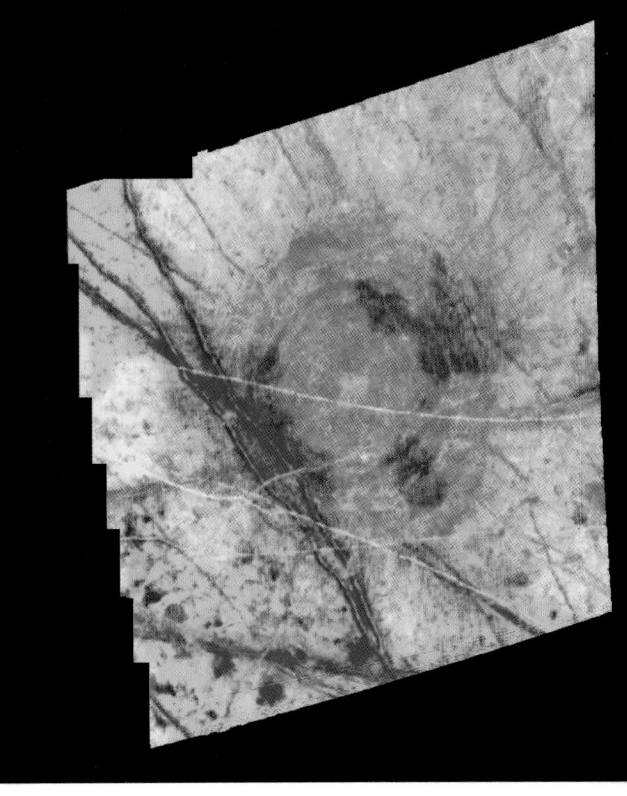

This composite image of part of Europa, shows the distribution of ice and minerals for the structure named Tyre. The image was created with data from Galileo's Solid State Imaging (SSI) camera and the Near Infrared Mapping Spectrometer (NIMS). Tyre, the circular feature, is 140 km in diameter (about the size of the island of Hawaii) and is thought to be the site where an asteroid or comet impacted Europa's ice crust. The blue in this image indicates areas with higher concentrations of mineral salts. These salts are similar in composition to those found in the bottom of Death Valley, California. The yellow-orange regions are areas that have a high surface abundance of water ice. The centre of this impact feature (located at 34° latitude and 146.5° longitude) appears to have a surface composed of coarse-grained ice.

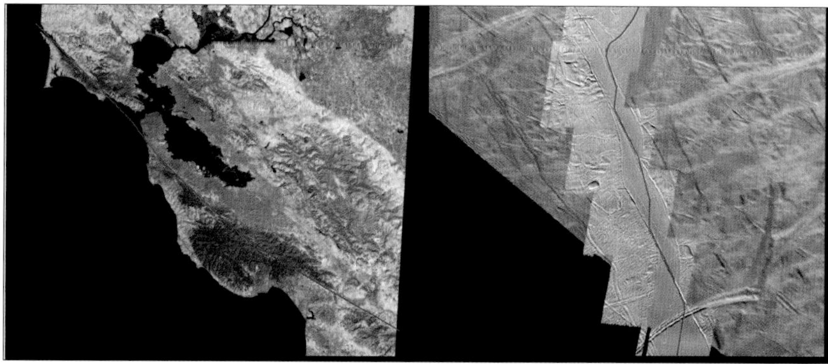

The mosaic on the right of the south polar region of Europa shows the northern 290 km of a strike-slip fault named Astypalaea Linea. The entire fault is about 810 km long, the size of the California portion of the San Andreas fault on Earth which runs from the California–Mexico border north to the San Francisco Bay. The left mosaic shows the portion of the San Andreas fault near San Francisco Bay that has been scaled to the same size and resolution as the Europa image. Each covers an area approximately 170 km × 193 km. The red line marks the once active central crack of the Europan fault (right) and the line of the San Andreas fault (left).

surface should rise and fall by several metres every few days, as Europa swings around Jupiter. If the crust is solid, then the tidal range may be no more than a few centimetres. In September 1999, Greg Hoppa, a University of Arizona scientist who had examined the Galileo images, concluded that he had found evidence for very large tides on Europa – tides that you could actually feel through the soles of your feet.

Hoppa could not try to measure the rise and fall of the surface directly, but he examined strange structures called flexi that cover Europa's surface. Flexi are curved cracks that appear as a series of arcs, joined together at each end to form a long, wavy crack across the surface, like the edge of a scallop shell. These cracks – unique to Europa – were first noticed in the Voyager images in 1979, but had defied explanation. Hoppa and his

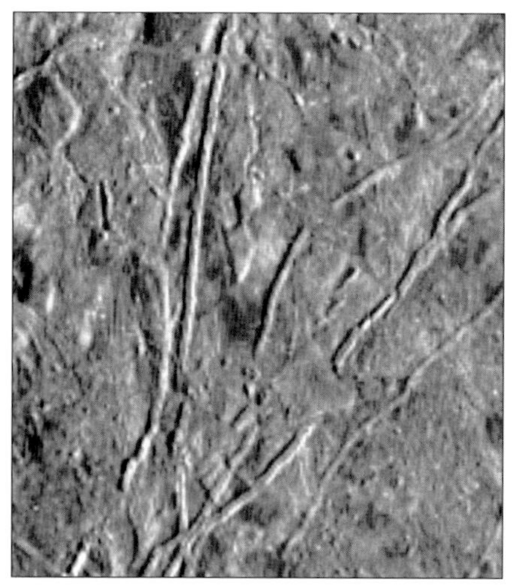

The area of Europa's surface shown here is about 124 km × 186 km across and shows features as small as a 800 m wide. Thick, tongue-shaped "lava" flows, the first seen on Europa or any of the icy satellites of Jupiter, are visible in several areas, including the lower-right quarter of the picture where one flow cuts across a prominent ridge. Most of the ridges on the left side of the picture appear to be partly buried or subdued by flows. The ice-rich surface of Europa suggests that the flows might also be ice, perhaps erupted onto the surface from the interior as viscous, glacier-like masses.

colleagues concluded that the flexi are in fact lines along which the thin ice crust of Europa is cracking and straining as the ocean beneath rises and falls. He says the tides of Europa rise and fall by as much as 30m – a range ten times greater than most terrestrial tides. "This causes Europa's ice shell to flex", Hoppa says. When the tidal stress exceeds the tensile strength of the ice, a crack forms. That crack propagates along a curved path on the surface until the stress drops below the strength of the ice, at which point the crack stops. Each arc in the flexi is 75–200 km long and forms over the course of three and a half days – the same time it takes Europa to orbit Jupiter. "You could probably walk along with the advancing tip of the crack as it was forming," Hoppa adds, "and while there is not enough atmosphere to carry sound, you would definitely feel the vibrations as it formed. What amazes me about this is just how long these features have been a mystery. We've been staring at pictures of them for twenty years, since Voyager. We didn't know what made them. And it seems that what they have been telling us all along is that an ocean was there when these things formed."

One puzzle remained, though. It was easy to see places where the ice crust of Europa was cracking apart, with new material welling up from the depths to fill the gaps. The triple bands seen crisscrossing Europa's surface are assumed to be analogous to the mid-oceanic ridges here on Earth, where magma from the mantle below the crust pushes its way up at a seafloor boundary between two plates, and pushes the plates apart. The plates that comprise the Earth's crust fit snugly together like a spherical jigsaw puzzle. So if they are moving apart at some plate boundaries, there must be places where they collide. This they do, crumpling together and sliding under each other, throwing up mountain chains and generating earthquakes and volcanoes. But where on Europa are ice "plates" colliding?

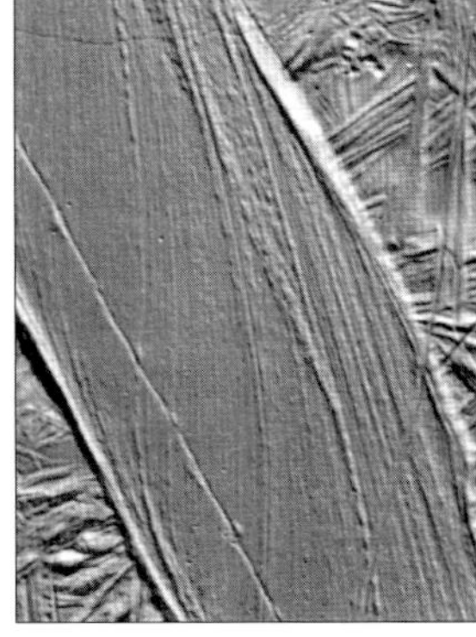

Like Earth, Europa has been an active world in the past. But is it still active today? This image shows evidence of past tectonic activity on the icy surface. The picture shows part of a grey band that formed as plates on the icy surface separated and material filled in the widening gap. North is to the top of the picture. In the centre of the image, a gently curving linear crack runs north to south and appears to be the location where the fault originally opened.

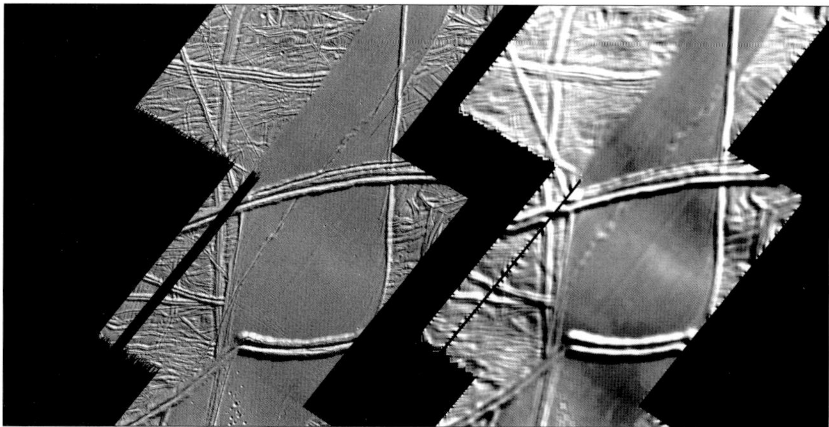

Folding on Europa. These images show ridges and troughs on the Europan surface caused by lateral compression of the ice.

The answer came in August 2000. Pappalardo and his former post-graduate student Louise Prockter, a British scientist who has spent most of her professional career examining Europa, announced that they had found folds on the surface. They say that Europa is not as flat as was thought, and the mountain-like features – found in three regions – are the first sign of compressional forces acting in the fractured Europan crust. "We learned from the Voyager images in the late 1970s that there was a lot of extension on Europa – that the surface was pulling apart and a slushy material was moving up through the gaps – but no one could find out how this new material was being accommodated", Prockter says. "Now we have finally found folds where the icy surface material compresses, and this will help us start to understand how Europa evolved and how it resurfaces." The two scientists, collaborating for a *Science* article, first noticed the folds in high-resolution Galileo images of the Astypalaea Linea fracture region. Near the large fracture zone they spotted fine-scale features – patterns of fractures and small ridges that mark adjacent crests and valleys – that characterize fold structures such as the Jura or Appalachian mountains on Earth.

The folds' orientation and location along Astypalaea Linea are consistent with the tidal stresses that scientists believe creates the pattern of large, canyon-like cracks on Europa's rotating surface. The fold crests are possibly tens to hundreds of metres high and spaced about 25 km apart. Pappalardo and Prockter spotted similar folds in two other regions, and

This is one of the highest-resolution images taken of Europa's Conamara Chaos region. Corrugated plateaus ending in icy cliffs over 100 m high dominate the top of this image. Debris piled at the base of the cliffs can be resolved down to blocks the size of a house. The fracture that runs horizontally across and just below the centre of the image is about the width of a motorway.

believe they could exist in other areas. One reason the folds have been hard to find is that Europa does a good job of hiding them: over time, the two researchers hypothesize, the folds "relax" (gradually flatten out) and push some material back into the interior for recycling.

Galileo had discovered a world of true mystery. Perhaps the most impressive pictures taken during the entire mission were the close-ups of Conamara Chaos. Some of these images show a fantastic world of towering ice cliffs several hundred metres high, littered with blocks of ice debris the size of a house. It is a landscape totally unlike anything seen in Earth's glacial regions. The Europan ice, as hard as rock and unsculpted by wind or rain, has piled into crazy formations that thrust high above the surface in Europa's feeble gravity. One day, astronauts will climb these cliffs of water.

So what is the current consensus on Europa's ocean? Nearly all scientists are agreed that under the ice lies liquid water. How much water, and how deep this ocean is, is a matter for debate. Bob Pappalardo is probably typical when he says he is "about 70 per cent sure there is a liquid ocean at depth today". Neal Ausman is more confident: "I think there's no question that there's an ocean down there."

Perhaps the most convincing evidence that what may turn out to be the biggest ocean in the Solar System lurks under the Europan ice comes not from Galileo's cameras but from its magnetometer, which is there to detect magnetic fields. Much to everyone's surprise, it had found that Gaspra and Ida both possessed magnetic fields, thought previously to be the preserve of major planets. At Europa, the magnetometer turned up even more of a surprise. Chief magnetometer scientist Margaret Kivelson explains how her instrument more or less proved the existence of the Europan ocean.

"In our first passes by Europa we discovered that there were changes in the magnetic field strength as we came in close, and as we travelled away. We could only account for this by assuming that Europa had a planetary-scale magnetic field much like the Earth's." She explained the complex mechanism by which Europa's magnetic field could be generated. The density figures for Europa precluded a large, conducting iron core. Some other conducting material had to be responsible – something that could interact with the massive magnetic field of Jupiter itself to create a small, transitory field that rose and fell as Europa orbited the giant planet.

"Why is there a time-varying field at Europa? It's because Europa is in orbit around Jupiter and Jupiter has a large magnetic field ... The magnetic field of Jupiter is substantially offset from the rotational pole. The result of that is that, as the planet rotates, the field rocks back and forth. This rocking field produced by Jupiter induces a field in Europa, and the best way of explaining this is to imagine a huge mass of salty water – a reasonably good conductor of electricity – under the surface.

Kivelson adds, "That kind of response can be produced in any conducting material. If it were happening in the deep interior of Europa it would not have produced a change in the field as large as what we measured with Galileo. So the alternative idea that the currents are flowing very near the surface was really the only way we could explain the kind of field we saw. Now, the next question is where are the currents flowing near the surface? It turns out that solid ice is not a very good conductor, even when it's not perfectly clean, so we don't believe that the currents that are necessary could flow in the surface ice layer. However, the temperature does increase as you move towards the centre away from the surface, possibly melting the ice. It doesn't take very much material of the nature of salt or acids to make liquid water a pretty good conductor."

Kivelson explains that only a global subsurface ocean will do: separate lakes of water under the ice could not account for the rise and fall of Europa's magnetic field as it swings around Jupiter. "We need a path which allows the currents to flow from equator to pole all the way around the body. Now that doesn't mean that the melt must be absolutely global, but there have to be connecting paths. It has to be

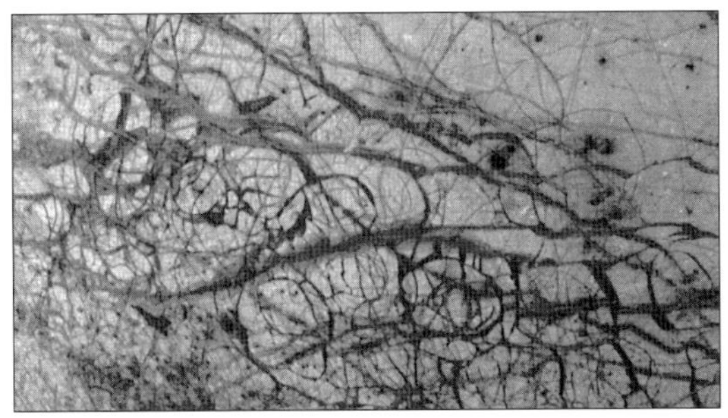

Europa's surface displays features in some areas resembling ice floes seen in Earth's polar seas. The ice crust has been severely fractured, as indicated by the dark linear, curved, and wedged-shaped bands seen here. These fractures have broken the crust into plates as large as 30 km (18.5 miles) across. Areas between the plates are filled with material that was probably icy slush contaminated with rocky debris.

possible to drive the current from equator to pole to equator to pole back to equator. But it doesn't have to be along a straight path – it can meander. So it is quite possible, for example, that if there is a region near the pole that is not melted, the currents could make a nice little detour round the region of unmelted material." In a paper published in *Science* in August 2000, Kivelson makes an even stronger case for a Europan ocean. Based on data from a flyby earlier in the year, she remarked that the test for an ocean just 7.5 km beneath the ice had been "passed with flying colours". An accompanying editorial concluded that the evidence is now conclusive. The Solar System has another waterworld.

Europan life?

An ocean on Europa could be the ideal home for life. As Galileo Imaging Team member Clark Chapman puts it, "The outer Solar System, beyond Mars and the asteroid belt, is indeed cold, only dimly warmed by the distant Sun. Basking as we do in the Sun's radiance and sustained

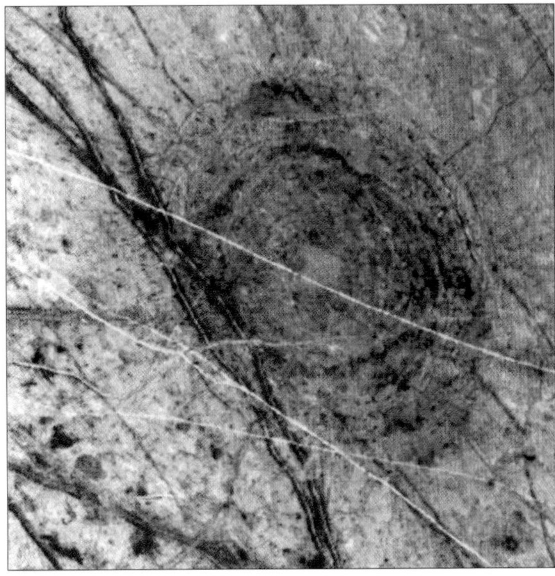

The 'bulls-eye' pattern appears to be a 140 km (86-mile) wide impact scar (about the size of the island of Hawaii) which formed as the surface fractured minutes after a mountain-sized asteroid or comet slammed into Europa. This approximately 214 km (132-mile) wide picture is the product of three images which have been processed in false colour to enhance shapes and compositions.

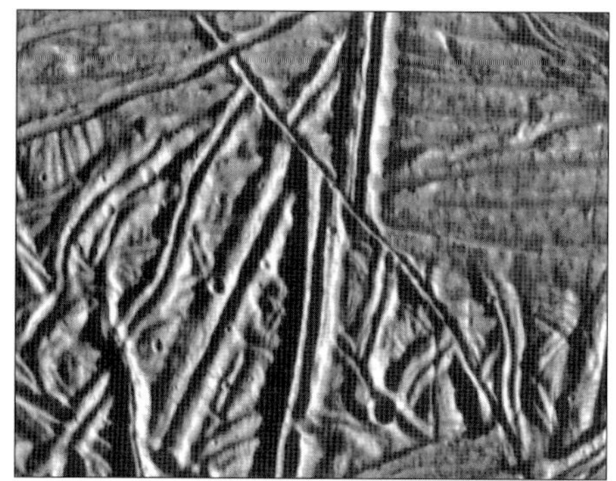

This photograph shows a Europan surface criss-crossed by multiple sets of ridges and fractures. The area covered by this image is approximately 15 × 12 km. The large ridge in the lower-right corner of the image is approximately 2.5 km across, and is one of the youngest features in this image, as it cuts across many of the other features.

by the ecology of photosynthesis, however, we are biased about the primacy of the Sun. Modern oceanography teaches us that we may be exceptional: teeming life exists on and beneath the ocean floor and may flourish in rocks beneath the Earth's surface. Energized by geothermal systems, life might even have originated below the Earth's surface, only later evolving upwards to take advantage of the ubiquitous sunlight."

As Galileo approached Jupiter, there was inevitably much speculation as to what the spacecraft might find, particularly at Europa. Some researchers even had the temerity to suggest that "fish" swam in the Europan seas – although most were content to fantasize about bacteria and algae. But even if the Europan ecosystem is limited to seafloor slime, that would still be a tremendous discovery.

We shall never know whether anything living lurks under the Europan ice until we drill through it, but one scientist has speculated that life may thrive on Europa much closer to the surface. Richard Greenberg of the University of Arizona thinks there might be an "ice forest" of plants and even animals waiting to be discovered just a few metres under the Europan surface. He says that green vegetation, living on sunlight and water, could fill the thousands of cracks that cover the surface of Europa. Greenberg points out that the tides on Europa constantly flex and bend the ice-shell crust, cracking it and allowing warm water to well up from underneath. Fissures about a metre across open up every three and a half days – the time it takes for Europa to orbit Jupiter and for the tide to rise and fall. These waters should be rich in chemical nutrients, he says, and the ideal place for life would be near the surface where sunlight would allow plants to grow. "That makes Europa extremely habitable." At the top of the cracks, the water boils and freezes into the vacuum of space. But a constant supply of warm water from below means that just a metre or so beneath the surface plants could flourish in a mixture of slush and liquid water. Here, sunlight can split molecules, producing chemicals that could supply the energy for life. And enough light will penetrate to a depth of 30 metres or so to drive photosynthesis – the process that "fuels" plants on Earth.

"You could have organisms a metre down, with roots hanging onto the cracks, unfolding their little leaves", Greenberg says. "If there were life in the cracks, the whole crack would be filled, it's such a hospitable setting." Some organisms might stay in one place, while others could ride up and down with the flow of water. As old cracks seal up after thousands of years, organisms would be forced to migrate to new cracks or evolve ways to survive when frozen. "It's a great set-up for advancement", he adds.

The future exploration of Europa

Chapter 12 looks at NASA's plans for exploring Europa, one of the most exciting places in the Solar System. The best-laid plans of science can of course be scuppered by bad luck and by Congressional funding committees, but all being well, in fifteen years or less we should know for sure whether Europa is the exciting world many people suspect. Under the surface there may slosh more water than in all our seven seas, filled with who knows what – sterile salty water, or a rich ecosystem to rival Earth's stunning coral reefs? Hopefully, the next generation of interplanetary probes to the Jovian system will answer all the questions scientists have about this extraordinary waterworld.

Europa was perhaps *the* highlight of the Galileo mission. Certainly the drama of finding a world that may harbour life has attracted a huge amount of public attention. But in the final months of its long tour, Galileo, now rather battered and nearing the end of its life, has made a series of spectacular discoveries about Jupiter's innermost moon – arguably the most dramatic and outlandish place in the Solar System. This is Io, the world of fire, to which Galileo returned in 1999 and 2000. We now journey on the second part of the extended mission to Io, bathed in radiation and flooded with searing lava curtains a kilometre high, and marvel at just how different a place can be to Galileo's home planet.

Io in natural colour – pretty much as the human eye would see it from the vantage-point of Galileo.

Chapter nine

FIREWORLD

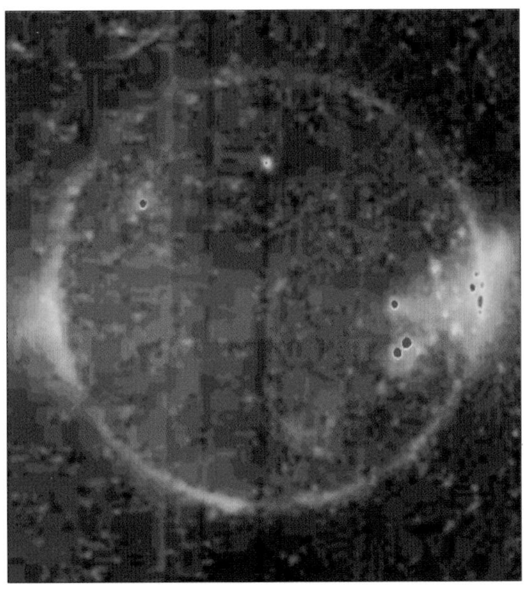

This image was acquired while Io was in eclipse (in Jupiter's shadow) during Galileo's eighth orbit, and shows the intense volcanism lighting up the night. The small red or yellow spots mark the sites of high-temperature magma erupting on to the surface in lava flows or lava lakes.

NOWHERE IN THE SOLAR System is quite like Io. Since the Voyager encounters of 1979, scientists have known that the innermost of Jupiter's moons is unique. No craters, no ancient impact basins; instead, a seething mass of boiling lavas, sulphur geysers 300 km high and volcanoes the size of Texas all dotted across a palette of crazy reds, bright greens and splurges of yellow. Looking something like a giant pizza covered with melted cheese and islands of tomato and ripe olives, Io is the most photogenic moon in the Solar System, and unquestionably the most active. For many, particularly in the Imaging Team, Io was the most exciting place that would be visited by Galileo, and they couldn't wait for the precious pictures and data to arrive.

Unfortunately the single planned flyby of Io, as the spacecraft made its initial approach to Jupiter before entering orbit in December 1995, did not go as planned. The problem with the tape recorder forced JPL into the decision to ignore Io on this first and quite probably only pass, concentrating instead on saving and returning the data from the atmospheric probe. Had it been any other moon this would not have mattered greatly; several flybys of Europa, for instance, were planned for the primary mission. But Io was different. Because its orbit takes it through the heart of Jupiter's hazardous radiation belts, Mission

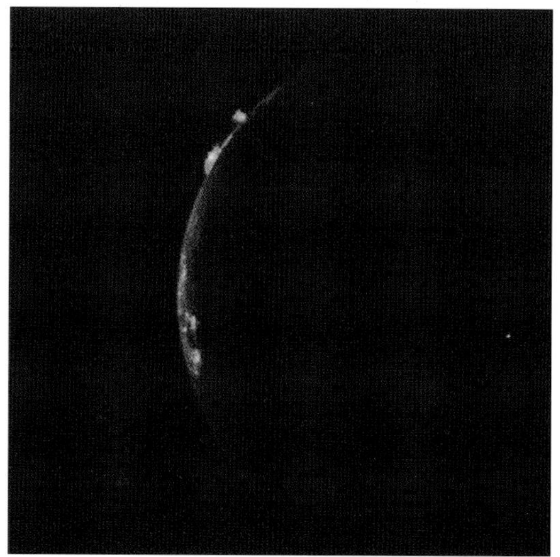

When this picture of Io was taken by Voyager in 1979 it nearly caused imaging scientist Linda Morabito to fall off her chair. The cause of the excitement? An erupting volcano, seen clearly in the top left of the picture as a plume ejecting into space.

Control decided that Galileo would risk only one visit to this moon, at the point where a close approach to Jupiter was unavoidable. After arrival day there were to be no more close-ups of this exciting place. The scientists hoping to get a better look at Io were devastated, angry and resentful. The primary mission, lasting until the end of 1997, would no longer include any close flybys of Io. However, there was still hope that Io's glories would be laid bare. No one seriously believed that Galileo would be junked at the end of the primary mission: all JPL spacecraft are designed around a very conservative life expectancy, at the end of which there should be a good chance of getting some more mileage out of them. As it turned out, Galileo was still in excellent health in December 1997, so an extended mission lasting a further two years or so was mounted. The extended mission would concentrate initially on Europa (and would return the spectacular results we have seen), and then, towards the end of 1999, the spacecraft returned to Io.

Fire and brimstone

Just before the Voyagers arrived for their historic Jupiter encounter in 1979, Stanton Peale, Patrick Cassen and Raymond Reynolds calculated that Jupiter's massive gravitational pull, combined with the tug from Europa and Ganymede, should be enough to generate huge tidal forces and a fantastic amount of heat in the little moon. Every time Io swept around Jupiter its crust should rise and fall by tens, maybe hundreds of metres. While its average orbit is circular, Io is in fact nudged this way and that by Europa and, to a lesser extent, Ganymede, as the moons pass one another – making Io's orbit elliptical to about one part in 250. This eccentricity is enough, on

Europa, to at least partially melt the ice crust and generate the Solar System's largest ocean. But on Io, much closer to Jupiter, the effect of tidal heating must be stronger still.

In fact Peale and his colleagues calculated that tidal heating might be sufficient to melt the crust of Io and generate volcanoes. Just a week after their findings – purely theoretical calculations – were published in March 1979, JPL navigation engineer Linda Morabito nearly fell off her chair when she saw what Voyager 1 had spotted – a geyser-like eruption spewing material hundreds of kilometres above the surface of Io. The discovery was a lucky one, for the "volcano shot" was captured only because the spacecraft's cameras were being used for navigational purposes. This is a strong candidate for the shortest time lag between scientific prediction and observational confirmation in the history of astronomy. The volcano was subsequently named Pele, and a second volcano, named Loki, was spotted in the same image.

Although Voyager spotted volcanoes on Io, the imaging team nearly didn't, because of some clever electronic trickery designed to make pretty pictures. Torrence Johnson was working on the Voyager mission when the pictures came in. He remembers, "Linda was working at one of these overexposed pictures and saw this faint thing sticking up from the side of it. She thought it might be something sitting behind it, a satellite or something; we didn't know what to make of it at first. So she took it to Brad Smith, who was the imaging team leader, and he recognized that it was probably a volcanic eruption. In fact they were debating as to how to deal with this discovery. When we all came into work on Monday morning, having gotten a rest after having been up for 72 hours during the encounters, we started working on the data. We had at the time sort of crude versions of the workstations we have now. We started systematically going through all the pictures and said, 'Well, is that the stuff that Brad is being so secretive about, that he doesn't want anyone to talk about?' So we grabbed Brad and asked him, and he says, 'Whoa – get that off the screen! No, wait a minute – that's another one!'

"What was amusing in retrospect is that the volcanoes weren't hard to see, and the only reason we didn't see them on the actual day of the encounter was that we got too clever for our own good. They were actually underexposed. We were very cautious; we kept the shutter tight. The computers clipped off all the plumes [erupting volcanoes], and so they were glaringly obvious in the raw data, but in these pretty pictures that we had fixed up to put on the TV screens, it was as if we'd known about them and tried to hide them. So the world's media would have discovered the volcanoes on Io if we hadn't tried to make the pictures look a little bit better."

As Voyager sent back more data it became clear that the two volcanoes picked up on that early image of Io were not alone. Its surface was positively seething with heat. Calculations showed that, on average, every square metre of the surface was releasing about 2.5 watts of warmth. By comparison, the most volcanic regions of Earth, such as the Wairakei area of New Zealand, Yellowstone Park or south-east Iceland, generate less than 2 watts per square metre. Io is much more volcanically active than our planet – up to then thought to be, geologically, the liveliest place in the Solar System. Io's size and bulk density are similar to those of our Moon, and suggest a composition predominantly of silicates – familiar rocks like basalt. Io was expected to be a cratered world like the Moon, but it clearly was rather more interesting.

As early as 1974, half a decade before the Voyagers arrived, spectroscopic analysis of Io from Earth showed the presence of sulphur on its surface. It was also known that Io lacked the surface water ice that was known to be present on the other Galileans. And just before the Voyagers turned up – at about the same time as Peale and his colleagues published their predictions that Io should be a seething hotbed of activity – telescopic observations recorded a temporary infrared brightening of Io, indicating that temperatures on at least part of its surface were momentarily reaching temperatures of several hundred degrees.

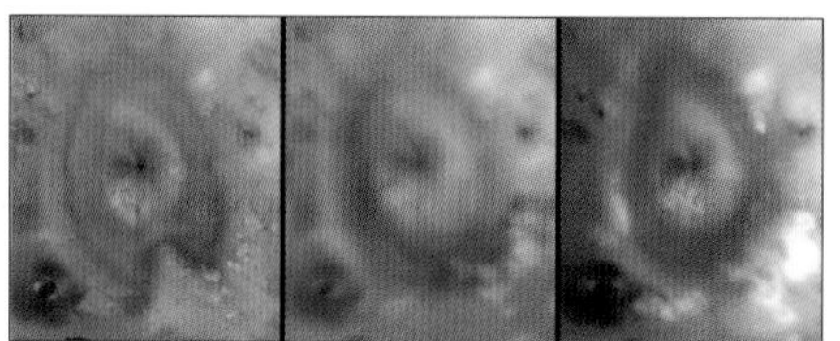

These three pictures reveal the changes around Pele, as seen by Voyager 1 (left), Voyager 2 (middle), and Galileo (right). The Voyager frames were taken in 1979 when the two spacecraft flew past Jupiter and Io. The Galileo view was obtained in June 1996. The dramatic differences demonstrate the volcanic activity of Io.

Fiery mystery

So Io is hot. Scientists wanted to know whether Io's volcanism resembled Earth's, or whether it was of some totally alien, Ionian, variety. Io was certainly newsworthy – the heat, the volcanoes and the sulphur immediately caught the public's imagination.

Here at last was a world that wasn't boring, dead and covered with craters like the Moon or Mercury. Venus might be interesting, but it is shrouded in clouds and you can't see a thing. Io's glories are laid bare for all to see. With its sulphur and fire, Io was turning out to resemble the underworld of ancient mythology. When the International Astronomical Union started naming Io's surface features, they soon hit upon a "theme": fire goddesses and gods, together with names of terrestrial volcanoes and mythological allusions to the underworld, now dot the maps of this world. Io has volcanoes called Loki and Pele, Prometheus and Marduk. (The old tradition of sticking to Latin for the nomenclature of the surface features of other worlds has long been abandoned, and charts of the outer Solar System bear names familiar to Polynesian, Native American and Hindu scholars.)

The geological rule-books would have to be rewritten when it came to Io. Here was a planet that was volcanic yet whose volcanism – apparently dominated by the element sulphur – seemed to be completely unrelated to the terrestrial variety. Here on Earth, volcanoes rely on supplies of liquefied silicate rock – lava – derived from the lower crust and upper mantle. Are Io's volcanoes purely sulphur-driven, or was the sulphur just a by-product of conventional

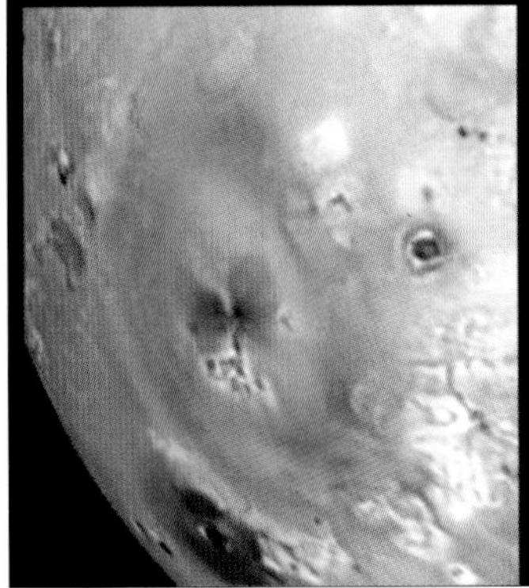
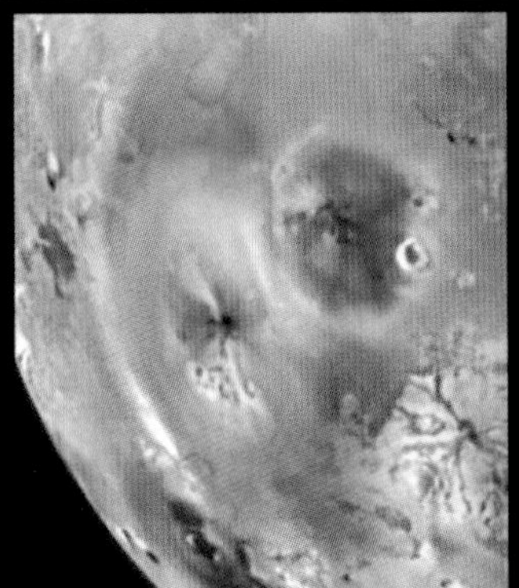

These images of Io show the results of a dramatic event that occurred on Io during a five-month period. The changes occurred between the time Galileo acquired the left frame, during its seventh orbit of Jupiter, and the right frame, during its tenth orbit. A new dark spot, 400 km in diameter, which is roughly the size of Arizona, surrounds a volcanic centre named Pillan Patera. Galileo imaged a 120 km-high plume erupting from this location during its ninth orbit . Pele, which produced the larger plume deposit southwest of Pillan, also appears different to its appearance during the seventh orbit, perhaps due to interaction between the two large plumes.

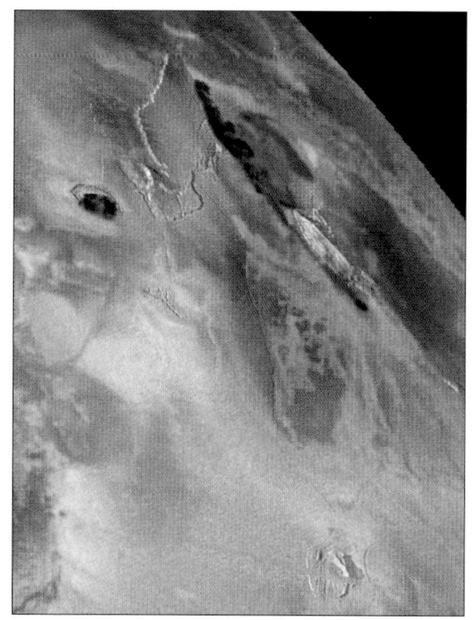

This image, comprising a mosaic of photographs taken throughout 1999, shows the Zal Patera region. The red colour to the south of the caldera is probably pure sulphur.

silicate volcanism as seen on Earth? Voyager's instruments were not sensitive enough to find out – but Galileo's were.

Back to Io

Even though the tape recorder was out of action for the 1995 flyby, some data were returned from Io. The spacecraft swept past the moon at a distance of just 900 km, and careful tracking of Galileo's altered velocity provided a very precise measurement of the moon's gravitational field. It seems incredible that simply by flying a space probe past a planet you can get a look at its insides, but it is true. If a planetary body like Io were homogenous (the same all the way through, from crust to core) the trajectory of an object in free fall as it moves past will be subtly different to the path it would follow if the body were composed of shells of different density, with a crust, mantle and core. The gravity-field data showed that Io has a very dense core of iron and perhaps iron pyrites – so maybe Io is full of fool's gold! Surrounding the core is a mantle of partially melted rock, topped by a very thin, mostly solid rock crust. This level of differentiation into layers makes Io more similar to, say, the Earth, than

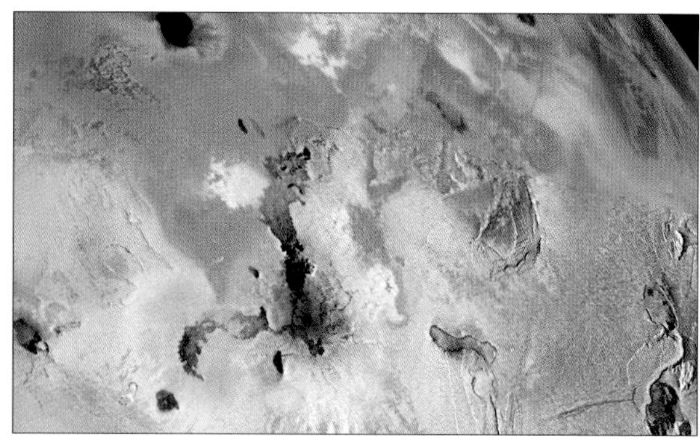

This picture of Io shows red material, which is often associated with areas where lava is erupting onto the surface and is thought to be a compound of sulphur, around the margin of Monan Patera (the elongated caldera just to the lower right of centre). The broad circle of bright, white material (just to the left of centre) is thought to be sulphur-dioxide which is being deposited from the plume Amirani.

This picture, taken on Galileo's ninth orbit, captures two erupting plumes on Io. On the limb of Io, in relief, is an eruption from Pillan Patera, a newly-active caldera. The other plume, in the centre of Io's lit side, is from Prometheus.

to the Moon or to Mercury, which are both fairly homogenous objects. It indicates that Io has been at least partially molten for long periods of time. Rather as shaking up a mixture of soil and water in a glass beaker, then leaving it to settle, the melting of a planetary body allows everything to separate out into layers, the heaviest at the centre and the lightest at the surface.

Nowhere else in the Solar System, at least today, do volcanic processes so completely dominate a planet's surface as on Io. As well as discovering the Pele eruption, Voyagers 1 and 2 found that, alone among the solid planets, Io has absolutely no impact craters at all. This is extraordinary – impact craters are ubiquitous. Mercury and the Moon are covered with them; Venus, Earth and Mars rather less so, thanks to the weathering, erosional processes and geological activity that reshape their surfaces. Even Europa has a few scars. Instead, Io is covered with hundreds of chasms and calderas – giant volcanic craters – gaping lava vents and blocky mountains. The total lack of impact craters implies that its surface is very, very young – nowhere on Io can be more than a million years old at the most, and many parts of the surface have been seen to change visibly over the course of two decades.

The Voyagers discovered about two hundred large calderas on Io's surface, ranging from 40 to 200 km across. Calderas look like circular

In February 2000, the camera spotted an eruption in progress at Tvashtar Catena, a chain of giant volcanic calderas. A dark, "L"-shaped lava flow marks the location of an eruption three months previously. A "fire fountain" of erupting lava can be seen in the top left of the image.

ulcers on Io's surface, and are filled with molten lava – either liquid rock or possibly liquid sulphur. One caldera, named Loki Patera, is bigger than the whole island of Hawaii. In addition to the calderas, the Voyagers discovered countless smaller volcanoes, and geyser-like structures responsible for spouting plumes of material into space.

In four years – the two years of the primary mission plus the bulk of the extended mission, which targeted mainly Europa – Galileo made two dozen orbits of Jupiter, concentrating on the other three Galilean satellites but giving Io a wide berth. Even monitoring from a distance was enough to yield important new insights into Io's surface, as well as producing the most stunning global snapshots of this world ever taken. For two years, Galileo was able to count volcanoes, analyse hotspots and watch for changes on the surface between each pass. Even from distances of 160,000 km or more it was able to spot significant differences on Io's surface from what had been seen by the Voyagers. In fact, such is the rate of change on the surface of Io that differences were apparent between one Voyager encounter and the next, a mere four months apart.

Then, in late 1999, JPL sent Galileo back to Io. All the science teams were ecstatic. The Galileo chiefs decided that, with most of its other scientific

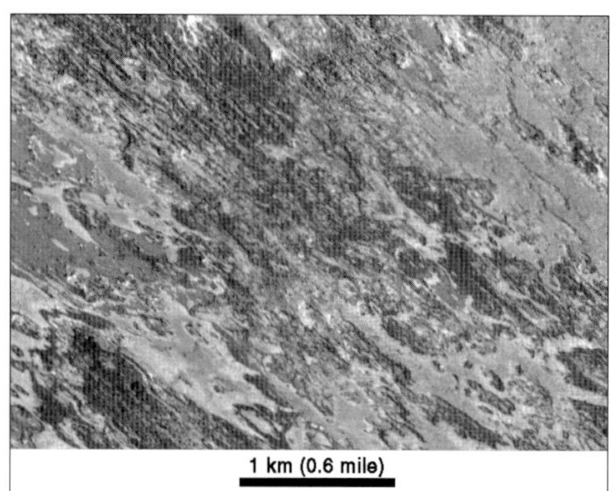

This image is the highest resolution image ever taken of Io. The resolution is 5.2 m per picture element. The bright areas are generally higher in elevation than adjacent dark areas. The surface appears to have been eroded by an unknown process, in places exposing layers of bright and dark material. Evaporation of solid ice may also play a role in separating the bright and dark materials.

This mosaic of images taken on Thanksgiving Day, 25 November 1999 shows a 1.5 km 1,300°C fountain of lava spewing above the surface of Io. In the original images, the active lava was hot enough to cause what the camera team describes as "bleeding" in Galileo's camera, overexposure caused when the camera's detector is so overloaded by the brightness of the target that electrons spill down across the detector. This showed up as a white blur in the original images. With the aid of computers, the whited-out area has been reconstructed to appear as it might without the "bleeding".

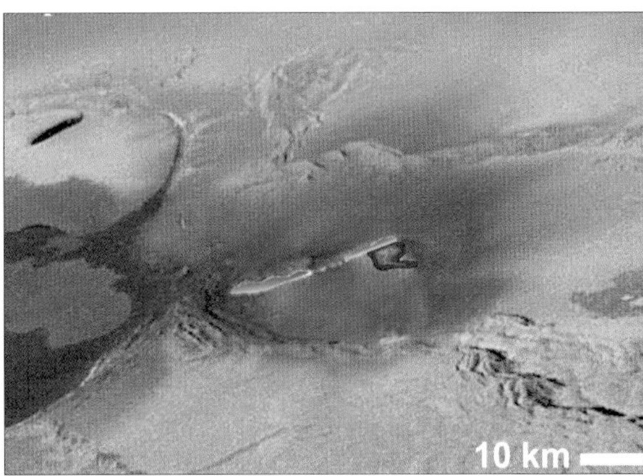

objectives addressed, there was little reason for the spacecraft to stay away from the volcanic moon any longer. "Of course, I was truly over the moon when I found out that we would do more flybys of Io", says Rosaly Lopes-Gautier, in charge of the NIMS team looking at this moon. "I knew it would be risky, and I didn't want to expect to get data and then go through the same disappointment as at the Jupiter arrival. My expectations were a lot more cautious." Torrence Johnson remembers that, "When Galileo finally returned to Io last fall [1999], the mission team was uncertain whether the spacecraft would survive the radiation. Several instruments suffered damage, but all continued to work and in the end returned spectacular data. Io's active volcanoes were finally captured up close and personal."

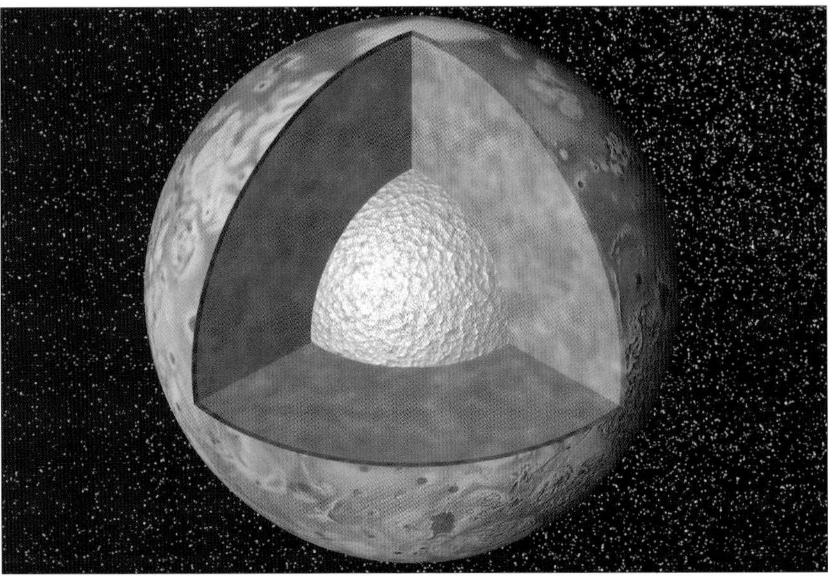

This model shows the interior of Io, as inferred from gravity and magnetic field experiments carried out by the Voyagers and Galileo. Io's iron-rich core can be seen surrounded by a silicate shell that extends to the surface.

As of January 2001, Galileo had made three close passes of Io – a week before the tenth anniversary of its launch, on 11 October 1999, on 26 November 1999 and on 22 February 2000 (designated I24, I25 and I27 respectively, the number denoting the Jupiter orbit since JOI, and the "I" that Io was the primary target). More than a hundred images were returned, some of spectacularly high resolution (about 10 m per pixel) and among the best from the entire mission. Galileo flew past Io at an altitude of just 300 km on one of its passes. In comparison, the best Voyager images had a resolution of around 500 m per pixel and were taken from hundreds of thousands of kilometres away. So this was Io up close and personal.

Hopefully the close-up flybys would solve the mystery of Io's volcanoes. Was it sulphur or silicate that accounted for most of the material ejected? Terrestrial lavas consist of molten silicate rock at typically 1,000°C, mixed with superheated water and gases, and rise to the surface through vents and cracks in the upper crust from magma chambers below. Volcanoes on Earth vary from place to place. The crust of our planet is divided up into a number of plates – about a dozen large ones and several more smaller platelets. Plate tectonics – the mechanism that governs the relative motion of these plates – explains the different types of volcanism found on Earth's surface. Broadly speaking, there are two types. Basaltic volcanoes such as those on Iceland and Hawaii ooze free-flowing molten basalt, a dark alkaline rock that contains very little quartz (silicon dioxide). These volcanoes are typically found where the crustal plates are being pulled apart, allowing molten rock to well up from deep below and fill the gap. They are also found (as on Hawaii) right in the middle of plates, above high-temperature plumes originating deep in the mantle of the Earth. As the plate drifts slowly overhead, the high-temperature plume acts as a blowtorch, punching a series of holes through the crust overhead and melting the rock. The result is a chain of volcanic islands. (Very long-lasting plumes may be responsible for the great Martian volcanoes of the Tharsis plateau, although geological analysis from orbiting probes has found puzzling amounts of acidic lavas at the top of these volcanoes. The lack of tectonic motion is what is assumed to be responsible for the great sizes of these volcanoes.)

This image, taken during Galileo's close flyby of Io on 25 November 1999 shows some of the curious mountains found there. By measuring the lengths of the shadows, Galileo scientists can estimate the height of the mountains. The mountain just left of the middle of the picture is 4,000 m high (comparable with the highest peaks in the Alps) and the small peak to the lower left is 1,600 m high – a little higher than Ben Nevis, the UK's highest peak. The image, centred at -8.1° latitude and 78.7° longitude, covers an area approximately 210 × 110 km.

The other type of terrestrial volcanism is found where tectonic plates collide, such as all around the Pacific Rim and the Mediterranean. Along the western seaboard of North America, the Pacific plate is being driven under the North American plate. Colliding plates tend to produce a very different kind of volcano: violent, with blocky viscous lava and huge quantities of gas and steam. While Mauna Loa on Hawaii is a quiet, predictable beast, silica and gas-rich "andesitic" volcanoes found along colliding plate boundaries are often anything but. Pompeii, Krakatoa, Mount St Helens and Pinatubo were all this type of eruption.

Because spectroscopic analysis had shown Io to be rich in sulphur, many people speculated that Io's volcanism consisted not of molten silicate lavas but of brimstone geysers, huge eruptions of molten sulphur spewing out over the surface and spattering the crust with yellow sulphur "snow". The other colours could easily be explained by sulphur's unusual chemistry. Unlike most elements, the solid form of this element exists in a number of different forms, known as allotropes, which crystallize at different temperatures. These allotropes are different colours – normal sulphur, the sort which is happy at room temperature, is bright yellow, but if molten sulphur is rapidly cooled from a high temperature, it solidifies into a reddish brown form. The

Galileo acquired the images in this mosaic of Hi-iaka Patera (the irregularly shaped, dark depression at the centre of the image) and two nearby mountains on 25 November 1999 during its 25th orbit. The sharp peak at the top of the image is about 11,000 m high, and the two elongated plateaus to the west and south of the caldera are both 3,500 m high.

Io has the tallest mountains in the Solar System, higher even than Earth's Himalayas or Mars' giant Mt Olympus. But Io's mountains have formed in a completely different way to those on Earth and Mars, which were uplifted along plate boundaries (Earth) or as the result of deep-rooted volcanism (Mars). It is thought that Io's giant mountains, some of which are nearly twice the height of Everest, may be a combined result of the heating, melting, and tilting of giant blocks of crust.

This 3D graphic of the Tohil Mons mountain was produced using Images taken with different lighting and from different positions by Galileo. It shows a topographic representation of what Tohil Mons looks like when seen from the northeast. The topography has been vertically exaggerated. The peak's height is about 6,000 m, plus or minus 200 m.

garish colours of Io could simply be explained by the presence of different sulphur allotropes on the surface, perhaps related to the distance from a central volcanic vent. Or perhaps impurities were mixed in with the sulphur. In addition, it is known that sulphur is a common element in the Solar System. We are not particularly aware of it, though it holds together the proteins in our bodies, and is also associated with Earthly volcanoes. Finally, the Voyagers found that the volcanism on Io happened at a temperature of just a few hundred degrees. This is far too cool for silicate volcanism – rocks like basalt and granite won't even soften at such temperatures, let alone liquefy.

On the other hand, many scientists weren't convinced that Io's volcanoes were brimstone vents. Sure, there was a lot of sulphur on the surface, but this could be a thin, superficial coating. In 1986, three years before Galileo was launched, a team of astronomers in Hawaii noticed a striking increase in Io's infrared brightness, and concluded that this was due to a massive eruption of at least 620°C, too hot for liquid sulphur. This was the first direct evidence that liquid silicates were at least partially responsible for Io's volcanism. When Galileo arrived at Jupiter, it started to monitor the surface of Io and found even higher temperatures – 1,600°C was typical of the brightest volcanic hotspots. The close-up pictures, when they started to arrive at the end of 1999, answered many of the long-standing questions about Io. To the scientists who had had to wait nearly four years for this moment it was a cause for celebration. As Rosaly Lopes-Gautier recalls, "I got the first data from I24 while on a conference trip to Italy. I went to Italy because the schedule said that I would not get my data until after I returned. In fact, the schedule changed. I knew I had data on the evening before giving my talk at this meeting. We had a banquet that night, and somehow I managed to drag three of my colleagues back to my hotel at midnight to work on retrieving the data. It was really comic. We were in Abano, a sleepy spa town, and I show up at the hotel around midnight, wearing a cocktail dress and with three men in tow. I got some very funny looks from the night clerk. Then we had to do a rewiring job in the room to make the modem work. We got our first results at three o'clock in the morning, huddled around a laptop in that Italian hotel room."

The pictures flooded in, and forced the planetary geologists to completely rethink their ideas of what a volcano should be. In close-up it was clear that Io was nothing like the Earth. Galileo saw huge mountains quite unlike anything seen on its home planet, or on Mars or Venus: massive blocks of rock, rising from the surrounding terrain in almost sheer cliffs thousands of metres high. In places, the blocks look like they have been tilted and twisted, almost mimicking the ice rafts of Europa. This proves that the bulk of Io's solid surface is composed of silicates, not sulphur; even in Io's low gravity, kilometre-high cliffs of solid sulphur would not be stable, and mountain ranges of the stuff would simply crumble away.

In some places, the lavas were hotter than anywhere seen on Earth. Forget low-temperature sulphur volcanoes – this was primeval volcanism not seen on the Earth for over three billion years, molten lavas at thousands of degrees. So-called ultramafic lavas containing molten pyroxene rock are thought to be responsible for many of Io's volcanoes – Lopes-Gautier calls them "primitive".

The calderas are even more dramatic than they were first believed to be. In the summer of 1999 Galileo flew over a chain of calderas named Tvashtar Catena and took a sequence of stunning colour and monochrome images. The molten lava was hot enough, and therefore bright enough, to overload Galileo's CCD camera. The lava producing fountains of fire shooting hundreds of metres into space. Galileo also took close-up images of Io's plumes, spouting sulphurous gas and dust hundreds of kilometres above the surface. These fountain-like structures are some of the most impressive sights in the Jovian system – indeed in the whole Solar System. Because of the low gravity (about a sixth of the Earth's, or about the same as that of our Moon) and the lack of any appreciable atmosphere, the plumes are able to reach great heights before falling back to the ground in an elegant and symmetrical mushroom shape. One of Galileo's most famous images, used on the cover of this book, shows a plume erupting about 150 km high over Pillan Patera, caught in perfect silhouette on Io's limb. These plumes can leave huge, multi-coloured welts on the surface – often concentric rings of yellow, red and gold. These cannot be explained by conventional volcanism, as calculations have showed that even the most powerful liquid-magma volcanoes could not spew material out of the surface with enough energy to reach the heights of these plumes. In addition, some of the plumes appear to migrate – following snakelike paths across the surface.

The Galileo scientists were ecstatic after looking at the pictures from the 1999 close passes. "The latest flyby has shown us gigantic lava flows and lava lakes, and towering, collapsing mountains", said Alfred McEwen of the University of Arizona. "Io makes Dante's Inferno seem like another day in paradise." He explained that ancient rocks on Earth and other rocky planets show evidence of immense volcanic eruptions similar to those seen on Io today. The last comparable lava eruption on Earth occurred 15 million years ago, and it has been over two billion years since lava as hot as that currently erupting on Io flowed on Earth. "No people were around to observe and document these past events", said Torrence Johnson when the pictures came in. "Io is the next best thing to travelling back in time to Earth's earlier years. It gives us an opportunity to watch, in action, phenomena long dead in the rest of the Solar System."

Scientists now believe that a number of processes are operating on Io to generate the outlandish volcanic landscapes seen by Galileo. First there is a lot of "conventional" silicate volcanism, molten rock erupting through the surface from below, either through single vents or along gashes. Sometimes whole slabs of crust are driven apart by magma, forming giant pools of liquid lava. The crust itself seems to consist of "islands" of silicate rock surrounded by "seas" of sulphur and sulphur compounds. Some of the plumes are thought to be the Ionian equivalent of terrestrial geysers. On our planet, geysers originate when water comes into contact with hot crustal rocks, causing it to become superheated and erupting to the surface. These are relatively small-scale phenomena, blowing water and steam to a height of a hundred metres at most. But on Io, with its low gravity, a geyser such as Old Faithful in Yellowstone National Park would rise to 30 km high!

On Io there is almost certainly no water. Most of its surface was baked or boiled dry aeons ago. Instead, the agents that drives Io's "geysers" and plumes are sulphur and sulphur dioxide. These eruptions are now thought to occur when hot silicate rock comes into contact with cold sulphur and sulphur compounds. The river-shaped structure emanating from the centre of the Prometheus plume can be interpreted as a stream of molten silicate lava which is coming into contact with a sulphur deposit on or just below the surface, causing it to instantly vaporize, sending a cloud of material up from below. Alternatively, liquid sulphur dioxide could come into contact with molten or very hot silicates just below the surface (in a process similar to that which creates the geysers on Earth), again flash-boiling and sending a plume of gas into space. In fact there seem to be several distinct types of plume, some (such as Prometheus) driven by liquid silicates, and some perhaps by liquid sulphur dioxide, producing what is known as "stealth" plumes – nearly invisible and composed of more or less pure sulphur dioxide vapour.

Prometheus is certainly one of the most interesting structures on Io, resembling a squeezed pustule running with dark blood. As Rosaly Lopes-Gautier and her colleagues put it in a 2000 *Science* paper on this puzzling place, "Prometheus of legend was bound to a stake, whereas Prometheus on Io has not been bound by any physical or mythological constraints." To explain this, they have proposed that Prometheus is formed by a very-high-temperature silicate lava stream meeting a frozen sulphur or sulphur dioxide "snowfield" below, causing it to flash into vapour. Since Voyager first glimpsed Prometheus two decades before Galileo, in 1979, the location of the eruption has moved 75–95 km to the west. The authors of the *Science* paper speculate that in 1979 the lava flow was at a higher elevation, on the edges of the caldera, and has since flowed downhill into a pre-existing hollow – or has melted its way into the crust. They conclude that, "By its own dynamics, Prometheus the wanderer may become Prometheus-bound."

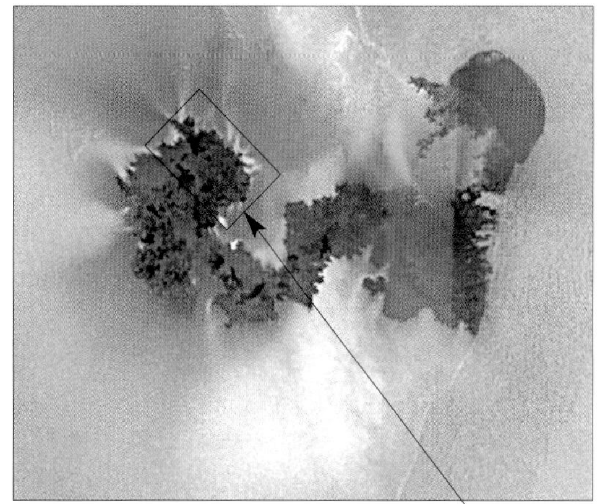

Prometheus is the "Old Faithful" of the many active volcanoes on Io. A broad, umbrella-shaped plume of gas and dust has been spotted above Prometheus by the Voyager and Galileo spacecraft every time the viewing conditions have been favourable. The volcano is surrounded by a prominent circular ring of bright sulphur dioxide apparently deposited by the plume. However, the origin of Prometheus' plume is a long-standing mystery: Where is the vent that is the source of all the gas and dust? Some clues are offered by this picture with a resolution of 170 m per picture element, which was taken by Galileo on 22 February 2000. To the right is a dark, semi-circular, lava-filled caldera. South lies a fissure from which dark lava has flowed towards the west (left). The lava flow extends 90 km from the source. Bright patches probably composed of sulphur dioxide can be seen in several places along the flow's margins. Galileo scientists are now studying whether heating of the volatile, sulphur dioxide-rich plains by encroaching hot lava might account for Prometheus.

This high-resolution picture shows a portion of the Prometheus area. The image has a resolution of 12 m per picture element. This entire area is under the active plume of Prometheus, which is constantly raining bright material. Hence, Galileo scientists interpret the darkest flows as being the most recent. They are not yet covered by bright plume fallout and perhaps too warm for bright gas rich in sulphur dioxide to condense. The older plains (upper right) are covered by ridges with an east–west trend. Bright streaks across the ridged plains emanate from the lava flow margins, perhaps where the hot lava vaporizes sulphur dioxide.

So it seems that Io is covered with both silicate volcanoes and erupting sulphur geysers. "In one sense, both sides in the sulphur/silicate wars were right", comments Torrence Johnson on Galileo's latest findings. "Sulphur allotropes may indeed be responsible for some of Galileo's garish colours. The Galileo camera is sensitive in the red part of the spectrum, where the Voyager cameras were blind, and its colour images of Io show strikingly red deposits around currently active volcanic areas. The evidence suggests that these fade back into the prevailing yellow-brown hue with time. Such traits are what we would expect if high-temperature sulphur allotropes were erupting on to the surface, then reverting to normal sulphur as they cool."

The burst of white light near Io's eastern equatorial edge is sunlight being scattered by the plume of the volcano Prometheus. Scattered light from Prometheus' plume and Io's lit crescent also causes the diffuse yellowish glow, which appears throughout much of the sky.

The skies of Io

Io's orbit takes it through a radiation maelstrom, a stormy sea of ions and magnetic lines of force. Jupiter has a vast magnetic field, generated by the massive sea of electrically conducting liquid metallic hydrogen in its interior. Jupiter's radiation belts are made up of charged particles – electrons and positively charged ions – trapped by the magnetic field. Earth's Van Allen radiation belts are composed of particles – protons mostly – from the solar wind that leak into the magnetosphere and get trapped. But Jupiter does not interact with the solar wind as much as the Earth does, thanks to its giant magnetic field which effectively diverts the stream of charged solar particles around it. Instead, Jupiter is surrounded by a powerful river of charged particles – in a doughnut shape called a torus – that are spluttered from Io's surface as the little moon circles the planet. Scientists have calculated that in this way Io loses about a tonne of surface material *every second*. However, even at this huge-sounding rate of mass loss, even a body the size of Io will have been whittled down by only a few hundred metres during the Sun's lifetime. Despite losing an apparently prodigious amount of material, Io is not being eroded away in front of our eyes.

Once in space, some of this material becomes ionized in Jupiter's rapidly rotating magnetic field, which from where Io is sitting sweeps past at about 60 km per second. This interaction creates a massive dynamo effect that was predicted as far back as the 1960s. Io has a thin atmosphere, as tenuous as a ghost, but nonetheless highly visible.

On 31 May 1998 Galileo took a picture of Io while the moon was in Jupiter's shadow. Gases above the satellite's surface produced an aurora – a ghostly glow of vivid reds, greens and violets. This phenomenon, produced by collisions between Io's atmospheric gases and energetic charged particles trapped in Jupiter's magnetic field, had not previously been observed to occur on a satellite. The green and red emissions are probably produced by mechanisms similar to those that generate auroral discharges in Earth's polar regions; the violet flashes mark the sites of dense plumes of volcanic vapour, and may be places where Io is electrically connected to Jupiter.

Galileo has also proved that Io is the source of the colossal quantities of dust being ejected from the region of Jupiter into the rest of the Solar System. In 2000, scientists, led by Amara Graps of the Max Planck Institute for Nuclear Physics in Heidelberg, Germany, analysed the frequency of dust impacts on Galileo's dust detector subsystem. They found peaks that coincided with the periods of Io's orbit – approximately 42 hours – and of Jupiter's rotation – approximately 10 hours. Although Io had been suspected as the source of the dust streams, it had been difficult to prove.

Unrivalled spectacle

An astronaut standing on the surface of Io would be privy to one of the most spectacular sights Solar System has to offer. Underfoot the visitor would crunch through a thick layer of frozen sulphur dioxide and perhaps gleaming crystals of red and yellow sulphur. On the horizon, a sheet of white-hot liquid erupts thousands of metres into the air, the molten lava freezing rapidly into dark rock particles before gently drifting back to the ground hours later. Magnificent geysers of sulphur blast into the sky. The landscape is primeval, a crazy mishmash of cliffs and canyons, and sheer walls of green-black olivine, the crystals glinting in the weak sunlight. Overhead, Jupiter glowers, its nightside aflash with lightning strokes generated by Io itself, a river of electricity connecting the giant planet to its little moon.

But there is little chance that a human will ever stand on Io – fascinating as it is, this world is simply too dangerous. For a start there are the volcanoes, which can erupt with no warning whatsoever. Against some of them, Earth's most active volcanoes are geological pussycats. Landing a manned spacecraft on a planet with 100-km geysers and 500-m fire curtains will not be easy. And to make things worse there is the radiation. Io is bathed in a sea of invisible danger so powerful that an astronaut wearing one of today's spacesuits would be fried in hours. Even Rosaly Lopes-Gautier, who has spent much of her life peering into terrestrial volcanoes, is not convinced that a visit to Io would be a good idea. She says, "I'd love to see Prometheus, the lava flow, the plume, and find out if our theory about how it forms is correct. But this is only in my wildest dreams, because I know that in reality the radiation would fry me pretty quickly. I'm a volcanologist, so Io is a kind of paradise for me – from a distance."

Torrence Johnson agrees that the view would be spectacular – even though the tourist would be fairly short-lived. "There is nothing anywhere in the Solar System that is anything like these geyser-like plumes. These things go up hundreds of kilometres, bringing up debris and probably sulphur dioxide snow. Now that would be very impressive!"

Galileo is sacrificing its electronic brain to the study of Io, its computer processors and memory chips subjected to the incessant fierce blasting from the radiation belts. Towards the end of 2000, JPL decided it would keep Galileo flying for as long as possible. Future encounters were planned for some of the other moons in an extended "Millennium Mission", but even Galileo's most enthusiastic supporters have had to accept that the time will come when the probe will – indeed must – be sacrificed. The alternative, to keep it flying, could result in a scientific and ecological tragedy beyond imagining.

Ganymede is the largest moon in the Solar System, larger than Mercury. Its ice surface is an enigmatic mix of dark and light terrains, which hint at a geologically active past.

Chapter ten

GANYMEDE AND CALLISTO – FROZEN SOLID?

IO AND EUROPA ARE active worlds. Squeezed by tidal forces, Io is a fiery cauldron of spewing brimstone and two-thousand-degree lavas; Europa's activity is less obvious, hidden beneath kilometres of ice. But things get a little quieter as we move away from Jupiter. A million kilometres from Jupiter's cloud-tops, the tidal forces exerted by the giant planet are greatly diminished. The worlds out here – giant Ganymede and battered Callisto – must rely on the heat of the Sun, and their own internal fires, for any activity. Here there are no volcanoes, no fire curtains and, at first glance, no oceans. However, on closer inspection these two worlds turn out to be rather more interesting than was thought.

Ganymede, Jupiter's third major satellite in order from the planet, is the largest moon in the Solar System and would easily be classed as a planet in its own right if it orbited the Sun. It is larger than both Mercury and Pluto, and three-quarters the diameter of Mars. But despite its size, Ganymede is a lightweight: its overall density of less than 2 grams per cubic centimetre means that it is probably made mostly of ice, in the form of a thick gravelly mantle and crust, surrounding a small rocky core. Its low density gives Ganymede a very low gravitational pull – less than that of our Moon's and just 15 per cent of Earth's.

The Voyager mission showed that Ganymede has some of the most bizarre topography in the Solar System. Unlike Europa, which at first glance looks like a smooth billiard ball, Ganymede's almost pure-ice landscape is a hotchpotch of mountains, channels and ridges, and what look intriguingly like lava flows. It has craters too, implying that much of the surface is ancient, but the concentration of these craters varies a lot from place to place. All this was known from the Voyagers, which took a series of images of this huge moon from a distance of several tens of thousands of kilometres. Planetary scientists were keen

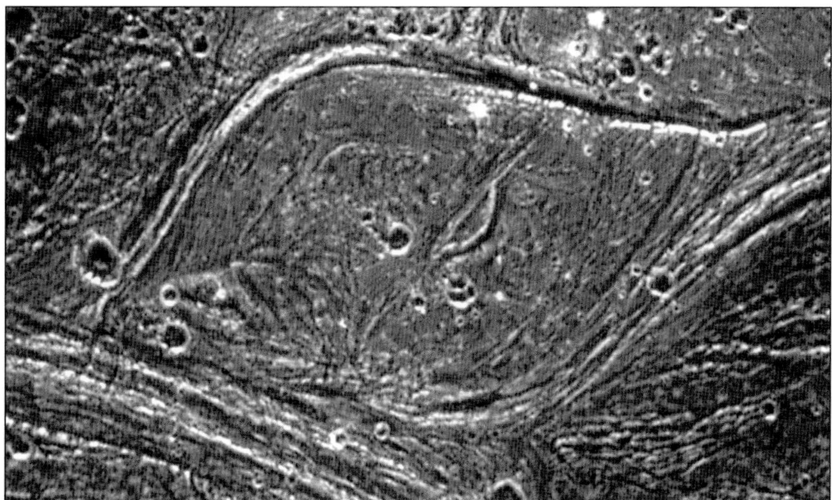

The 80 km-wide lens-shaped feature in the centre of the image is located at 32° latitude and 188° longitude on Ganymede along the border of a region of ancient dark terrain known as Marius Regio, and is near an area of younger bright terrain named Nippur Sulcus. The lens-like appearance of this feature is probably due to a shearing of the surface, where areas have slid past each other and also rotated slightly.

to know more about Ganymede, and they did not have to wait long into the Galileo mission to do so – Ganymede was to be the first "proper" encounter made by the spacecraft after its arrival at Jupiter and the downgraded Io flyby.

"Ganymede One" (G1), the first of Galileo's close-up encounters with a Jovian moon that would result in pictures, was to be the first test of the new low-gain tape recorder system perfected by JPL after the problem with the High Gain Antenna became apparent. The system had already worked well when called upon to play back the data from the atmospheric probe, and the particles and fields data gathered during JO1 and the "blind" Io flyby. But taking pictures and sending them home is a far more "bandwidth hungry" process than transmitting numerical data. The Galileo team would have to pick and choose where they pointed the camera as it swung past at just 845 km on 27 June 1996.

The original plan had been for Galileo to take hundreds of pictures while flying past Ganymede, but this would no longer be possible. Instead, it was decided to acquire a few images from each of the two basic types of terrain known on Ganymede – classified very broadly

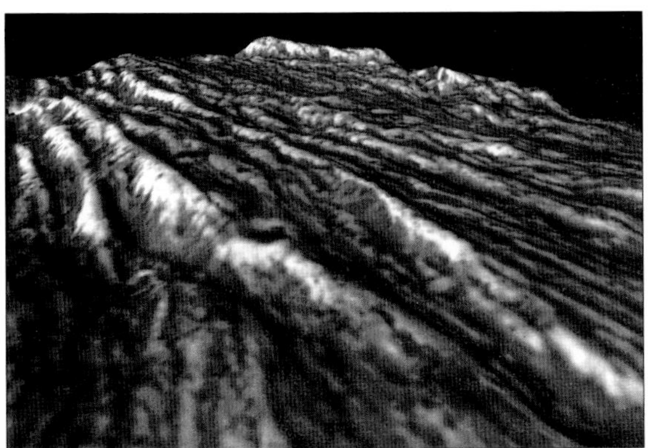
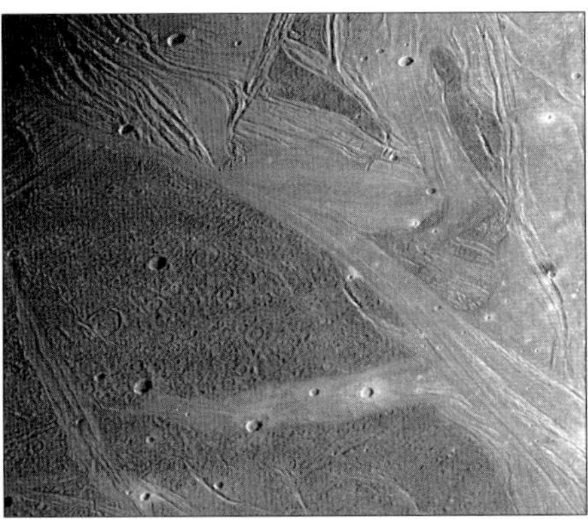

LEFT This image is a computer-generated perspective view of ridges in the Uruk Sulcus region of Ganymede. This area is part of the bright-grooved terrain that covers over half of the moon's surface, where the icy surface has been fractured and broken into many parallel ridges and troughs. Bright icy material is exposed in the crests of the ridges, while dark material has collected in low areas. RIGHT View of the Marius Regio and Nippur Sulcus area of Ganymede showing the dark and bright-grooved terrain which is typical of this satellite.

as "light" and "dark" terrain. Some of the target areas contained long ridges that looked for all the world like places where the crust of Ganymede had been rent asunder. Torrence Johnson said at the time, "The first images will show areas on Ganymede that have been tectonically ripped apart – the high-resolution pictures should tell us a lot more about the forces that have shaped Ganymede's crust." The flyby was also the first opportunity to get a close-up look at Jupiter itself; Galileo was to fly right over the heart of the Great Red Spot. Unfortunately a technical glitch – the most serious since the tape recorder incident – meant that the camera failed to home in on the centre of the Spot.

The results – including the pictures – from G1 would not arrive for two weeks. It was a frustrating wait for the JPL team and the scientists, a wait they would just have to get used to, as it would be repeated at every subsequent encounter. Finally, in early July 1996, the first pictures came rolling in. "It's been like Christmas every day", said Brown University scientist Jim Head at the time. "As we come in we open another packet from Ganymede." "Remarkable, absolutely stunning" was mission chief Bill O'Neil's comment. After all the delays, after all the glitches and near-disasters, Galileo was finally showing what it could do.

The pictures "exceeded our wildest expectations", according to camera chief Mike Belton, happy at last to have some images to play with after the loss of the Io pictures. This giant moon was now revealed in close-up for the first time. The region known as Uruk Sulcus looked as though someone had been over it with a giant rake, leaving long parallel grooves that have no analogue anywhere else in the Solar System. Other areas showed evidence of faulting and folding, with great sheared blocks of ice seemingly stacked up at random, embedded in a frozen matrix. This all pointed to an active world, one of great crustal movements, where icy plates are crushed together in some places, torn apart in others.

Clearly, Ganymede was once very active, although the density of craters on its surface – especially in the darker areas – implies that the forces that have shaped its crust have not operated for some time, possibly for billions of years. Arcing across the dark terrain are long depressions called furrows, which are troughs about 5–10 km across. Many of these furrows trace out concentric rings, and are probably "ripples" from colossal impacts early in Ganymede's history. Four and a half billion years ago Ganymede's interior still glowed with radioactive fire, and warm slushy ice – or even liquid water – lay just 10 km below the surface. Occasionally – probably every few thousand years – one of the mountain-sized chunks of space debris that littered the outer Solar System would impact Ganymede at a speed of 50,000 kph or more. The resulting explosion blasted through the relatively thin brittle crust. Slush and water then flowed into the central crater, dragging along giant slabs of solid ice. Kilometre-high waves of pulverized ice and water then spread out, freezing in the concentric rings we see today.

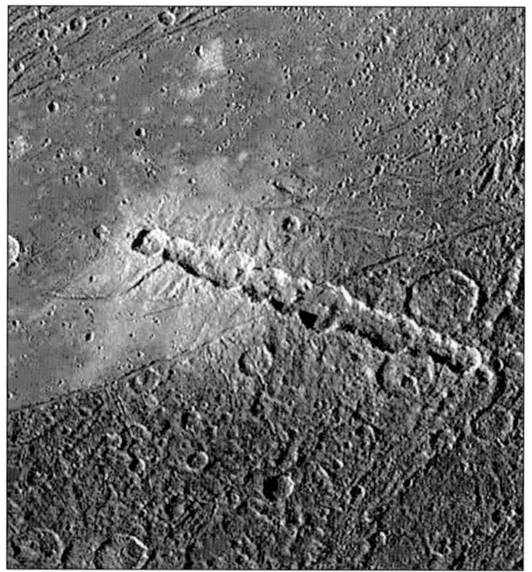

This chain of craters on Ganymede has been named Enki Catena. They probably formed by a comet, which was pulled into pieces by Jupiter's gravity as it passed too close to the planet. Soon after this break-up, the 13 fragments crashed on to Ganymede in rapid succession.

A growing moon?

The grooves and faults that scar Ganymede's surface are testament to a violent past. Like everywhere else in the vicinity, Ganymede was battered with space debris. But the surface of Ganymede appears to have been sculpted mainly by internal forces similar in many respects to the tectonism that governs Earth's geology. If either collisions or extensions predominated, our planet would be either shrinking or growing, and that does not seem to be happening. On Ganymede there is little evidence of collisional features, but plenty of evidence for extension – implying that at some point in its history the entire moon has expanded.

How could this happen? One possibility is that Ganymede's ice – which accounts for a fair bulk of the moon – underwent what is called a phase change at some time in its history. Ice (and some other substances, such as sulphur) can crystallize in different ways, depending on the temperature at which the solid was formed. "Warm" ice, the ice we are familiar with, crystallizes into hexagonal crystals; ice formed at much colder temperatures crystallizes into cubes. Each form takes up a different amount of space, and it is possible that at some time in its history Ganymede's ice mutated into a new form, growing as it did so while dropping in density.

Ganymede's magnetism

Galileo did not just uncover some remarkable cryogenic geology at Ganymede. The space physics instruments, including the magnetometer, discovered that Ganymede has its own magnetic field – the first to be found surrounding a moon. "What we've found is a magnetosphere within a magnetosphere", said Torrence Johnson. "While we expected some degree of interaction between Ganymede and Jupiter's magnetic environment, the size and the effect at Ganymede was completely unexpected. We knew Ganymede was an interesting place. What we have just found makes it even more exciting."

Galileo returned to Ganymede a few months later, on 6 September, for its second flyby. Data from the two passes, published in *Nature* in December 1996, confirmed that Ganymede is rather less dead than had been believed. Using extremely precise measurements of the spacecraft's trajectory, investigators on Galileo's celestial mechanics team have also been able to confirm that Ganymede's interior is differentiated, probably having a three-layer structure. "The data shows clearly that Ganymede has differentiated into a core and mantle, which is in turn enclosed by an ice shell", said JPL planetary scientist John Anderson, team leader on the Galileo radio science experiment. "Combined with the discovery of an intrinsic magnetic field, our gravity results indicate that Ganymede has a metallic core about 250 to 800 miles [400 to 1,300 km] below the surface. This is surrounded by a rocky silicate mantle, which is in turn enclosed by an ice shell about 500 miles [800 km] thick. Depending on whether the core is pure iron or an alloy of iron and iron sulphide, it could account for as little as 1.4 per cent or as much as one-third of the total mass of Ganymede."

Margaret Kivelson says that the presence of a magnetic field can tell us as much about a planetary body as a close up photograph – maybe more, as a photo is only skin-deep. "I have been interested in the question of whether there were any signs of magnetic fields on all of the moons of Jupiter", she says.

During the next four years, Galileo flew by Ganymede another half-dozen times. Kivelson says we still don't know exactly what is going on inside, but there are some hints nonetheless. "The question is, does Ganymede have any action beneath the surface? Its composition is much like Europa's, with a metallic core, a rocky mantle and an outer ice layer. But the remarkable thing about Ganymede is that it has very clearly a permanent, unchanging magnetic moment, very much like that of Earth, and we believe it's produced by currents driven in a fluid outer core."

She went on to hint at something even more exciting. Ganymede might have not one but two magnetic fields: one intrinsic, produced by an iron core; the other transitory, produced by a vast sub-surface salty ocean swirling through Jupiter's magnetosphere. Even if this were so, the signal from the latter field will be very weak, and hard to detect. "Just because you have that kind of field present, that doesn't preclude having an inductive field produced by currents flowing in a near-surface ocean. What it does do, however, is make it very hard to look

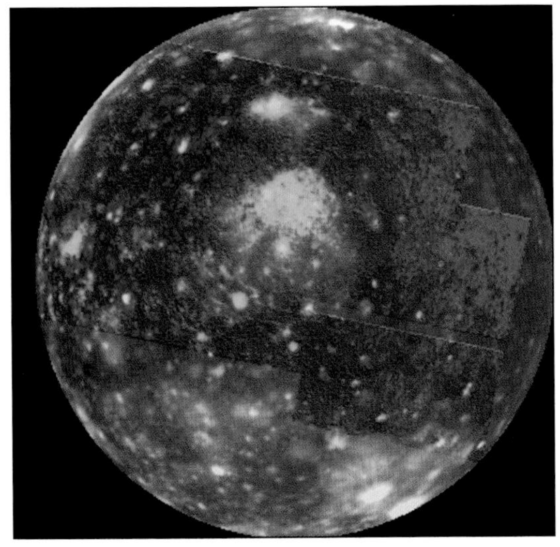

Callisto is Jupiter's third major moon, and was originally thought to be the least interesting of the quartet. Magnetometer data from Galileo have revealed, however, that Callisto could be home to a vast under-ice ocean, just like Europa.

for the small changes that such a process would produce."

Ganymede does not generate enough tidal heating in its orbit around Jupiter to melt a substantial layer of ice close to the surface. But there could be a water layer there nonetheless – hundreds of kilometres down, kept warm not by gravitational squeezing but by residual radioactive heat from Ganymede's interior. There is also evidence that Ganymede's orbit in the past may have been much more eccentric (i.e. elliptical), which would have periodically brought it close enough to Jupiter to generate a lot of tidal heating. This could be the key to the "fossil" tectonism seen on Ganymede's surface today. Even if the moon is still hot inside, in terms of geological activity it is now a shadow of its former self. The forces that shaped its crust simply do not operate today; the present surface features preserve a record of a more youthful, livelier age. Ganymede may be dead at its surface, lacking the excitement of Io and Europa, but compared with the outermost of the Galilean satellites, it looks positively lively.

Callisto – dull but with hidden depths?

"Callisto looked very boring", says Kivelson, summing up the attractions of this pockmarked moon. The outermost of the Galilean family, Callisto seemed from the Voyager photos to be the least interesting of the four. Covered in craters, it resembles our Moon, or Mercury. No volcanoes, no ice flows, no mountain-building either now or in the ancient past. Callisto's surface, the most heavily cratered landscape in the Solar System, presents a picture of a world unchanged for four billion years. It is not surprising that Callisto lacks the dramatic landscapes of the three innermost Galileans. It is the only one of the four main moons that is not "entangled" in gravitational resonance – with a circular orbit, its crust is not kneaded and squeezed by the interaction of Jupiter, itself and its neighbours.

Callisto even lacks large mountains. Our Moon, which is almost as heavily cratered, at least has some pretty dramatic relief to liven up the landscape. On Callisto, the fact that the surface is largely of water ice means that any relief created by a disturbance tends to "relax" – to flow downhill over time. Four enormous ring-shaped impact basins are found on Callisto. The largest, named Valhalla, has a bright central region 600 km in diameter, surrounded by concentric rings, the largest of which is 3,000 km across. The second-largest impact basin, named Asgard in accordance with the Norse theme, measures about 1,600 km across.

Half-baked

As Galileo sent back more data from Callisto, a picture of a rather more interesting world began to be revealed. Bob Pappalardo sums up the new picture of this moon: "Callisto has been facetiously called the boring Galilean satellite because images show it to be a heavily cratered world with scant signs of faulting or cryovolcanic resurfacing that would signal past geological activity. Thanks to Galileo, we now realize that Callisto is far from boring – in fact it is quite puzzling and mysterious in many ways." Scientists now believe that Callisto is different from Io, Ganymede and Europa, which all have differentiated structures. There is strong evidence that Ganymede is separated into a metallic core, a rocky mantle and an ice-rich outer shell, while Io has a metallic core and a rock mantle but no ice. Europa appears to be similar, albeit with a thin ice layer. Because Io, Ganymede and Europa are closer to Jupiter than is Callisto, they have been more affected by gravitational squeezing and subsequent heating. Over time, the forces exerted on the three inner Galileans have caused different constituents such as water ice, rock and metal to separate into different layers. Callisto, however, nearly 2 million km from Jupiter, has not been tidally heated and is "half-baked" compared with the other large moons, its ingredients having just begun to separate out, but still largely mixed together – like a fruitcake.

The image on the right shows one of the largest impact structures on Callisto, the Asgard multi-ring structure. Asgard – shown in close-up to the left – is approximately 1700 km across and consists of a bright central zone surrounded by discontinuous rings.

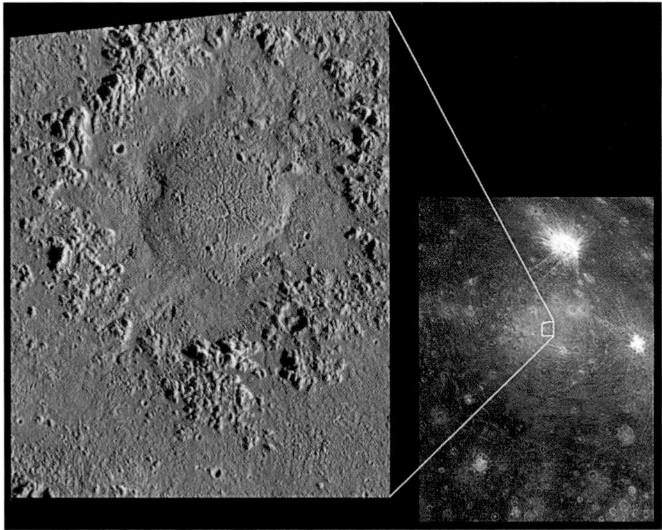

Surprise finding

At the end of the Primary Mission, then, Callisto was thought to be a dead, homogenous mixture of ice and rock, the silicates perhaps embedded in the frozen water like currants in a cake. Callisto had never been hot enough, for long enough, to differentiate into layers as the other Galileans had. But in October 1998 it was claimed that Callisto, of all places, may have a sub-surface ocean! Margaret Kivelson said at the time, "We thought Callisto was just a hunk of rock and ice. The new data certainly suggests that something is hidden below Callisto's surface, and that something may very well be a salty ocean."

Her claim was inspired by the Galileo magnetometer data from Europa, which seemed to point to a deep ocean of salty water just below the surface. Kivelson and her UCLA colleagues set out to test a similar theory about Callisto, "although it seemed far-fetched at the time", she said. The team went back and studied data obtained during Galileo's flybys of Callisto in November 1996, and June and September 1997. Kivelson and her colleagues found signs that Callisto's magnetic field, like Europa's, is variable, which can be explained by the presence of varying electrical currents associated with Jupiter that flow near Callisto's surface. Their next challenge was to find the source of the currents. "Because Callisto's atmosphere is extremely tenuous and lacking in charged particles, it would not be sufficient to generate Callisto's magnetic field, nor would Callisto's icy crust be a good conductor, but there could very well be a layer of melted ice underneath. If this liquid were salty, like Earth's oceans, it could carry sufficient electrical current to produce the magnetic field."

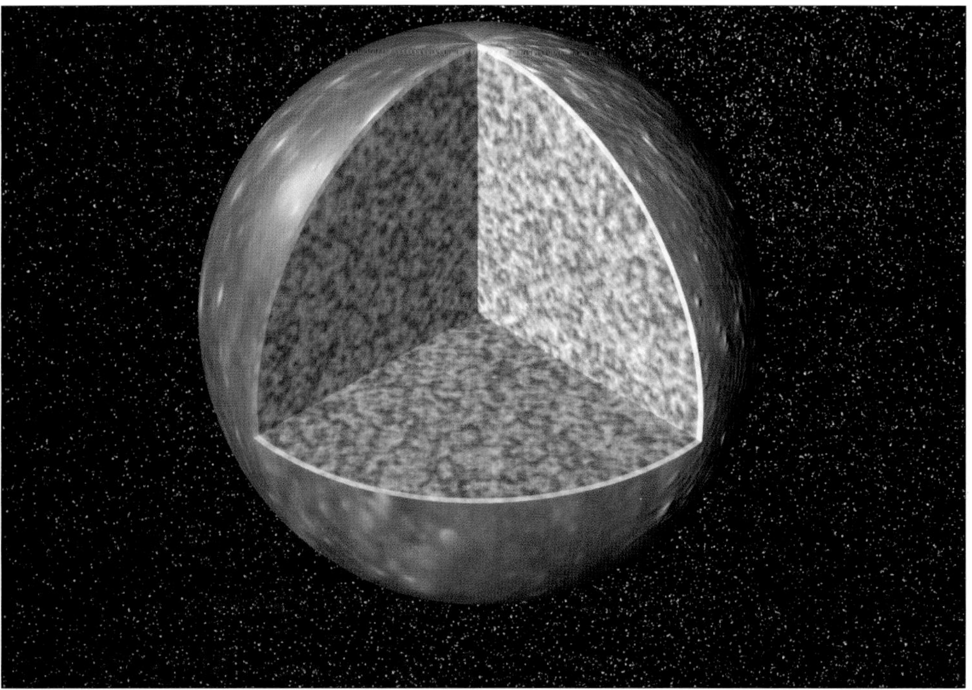

The interior structure of Callisto is still not entirely understood. It is likely that this moon consists mostly of an undifferentiated "porridge" of ice and rock, perhaps capped by a water ocean/ice crust.

This was extraordinary. The dullest world in the Solar System was turning out to be one of the most interesting. Callisto had kept its secrets well hidden, but now it seemed that underneath all those craters there could be an ocean and, who knows, even life. Torrence Johnson urges caution on this last point, however: "The basic ingredients of life – what we call prebiotic chemistry – are abundant in many Solar System objects, such as comets, asteroids and icy moons. Biologists believe that liquid water and energy are then needed to actually support life, so it's exciting to find another place where we might have liquid water. But energy is another matter, and currently Callisto's ocean is only being heated by radioactive elements, whereas Europa has tidal energy as well, from its greater proximity to Jupiter."

The idea of a Callisto ocean came as a complete shock. As Margaret Kivelson admitted, "This was a total surprise and still is. Regard it as a mystery. We were just very lucky that the first two passes occurred at a time when Jupiter's magnetic field was rather quiet and stable."

This mosaic of two images shows an area within the Valhalla region on Callisto. North is to the top of the mosaic and the Sun illuminates the surface from the left. The smallest details that can be seen in this picture are knobs and small impact craters about 155 m across. The mosaic covers an area approximately 33 km across.

The minor moons of Jupiter

There is more to the Jovian system than a gas giant and four big moons. But it took a long time to find this out. It was not until 1892 that E.E. Barnard of the Lick Observatory discovered a fifth Jovian moon, circling the planet every 12 hours, which he christened Amalthea. The Voyagers showed it to be a 270-km-long, vaguely egg-shaped lump of rock. Since then more tiny satellites have been discovered, telescopically from Earth and by the Voyager space probes. Most are only a few kilometres across, and all are irregularly shaped. Jupiter also has a tenuous ring system, invisible from Earth, first glimpsed by the Voyagers and captured by Galileo in some beautiful photographs. The Voyagers first revealed the structure of Jupiter's rings: a flattened main ring and an inner, cloud-like ring, called the halo, both composed of small, dark particles. One Voyager image seemed to indicate a third, faint outer ring. The Galileo data reveal that this third ring, known as the gossamer ring because of its transparency, consists of two rings. One is embedded within the other, and both are composed of microscopic debris from two small moons, Amalthea and Thebe.

A total of 24 minor moons are now known, four within the orbit of Io, and 20 beyond Callisto. Eleven of these were discovered towards the end of 2000 by astronomers at the University of Hawaii. One of the 11 new satellites, designated S/2000 J1, turns out to have been a previously observed but long-lost object, known originally as S/1975 J1. Galileo has managed to take pictures of some of the inner moons, but little is known of the outer satellites as yet – there may even be more to be discovered.

The four innermost moons are known as ring moons, because they supply the debris that form Jupiter's ring systems. All are reddish in colour, spattered as they are with sulphur spewed out from Io's volcanoes. They are, in order from Jupiter:

> **Metis**, 40 km in diameter, 128,000 km from Jupiter. This moon orbits within the Main Ring.
> **Adrastea**, 20 km in diameter, orbiting 129,000 km from Jupiter. This moon, the smallest of the four innermost satellites, skims along the Main Ring's outer edge.
> **Amalthea**, 270 km long, orbiting 181,300 km from Jupiter. This is the largest of the four inner moons, and lies at the outer periphery of the inner Gossamer Ring.
> **Thebe**, 100 km in diameter, orbiting 222,000 km from Jupiter. Thebe is near the outer periphery of the outer Gossamer Ring.

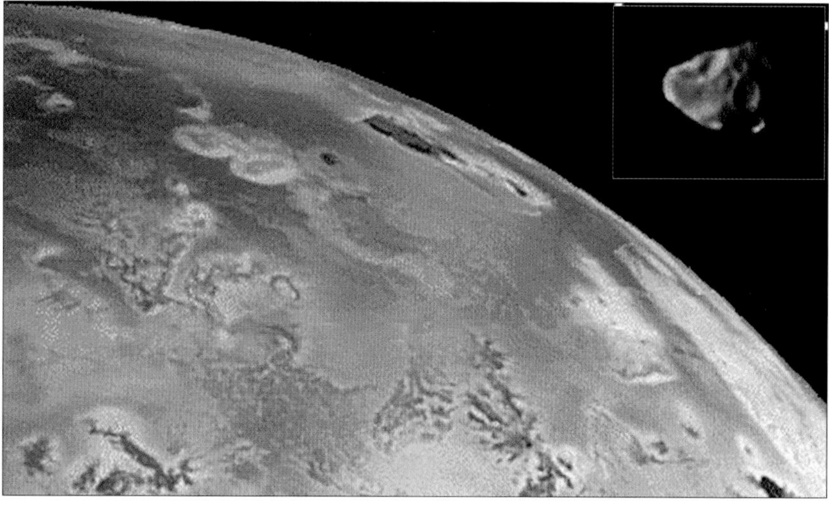

This composite image shows Jupiter's moon Io compared with Amalthea, one of the innermost "moonlets".

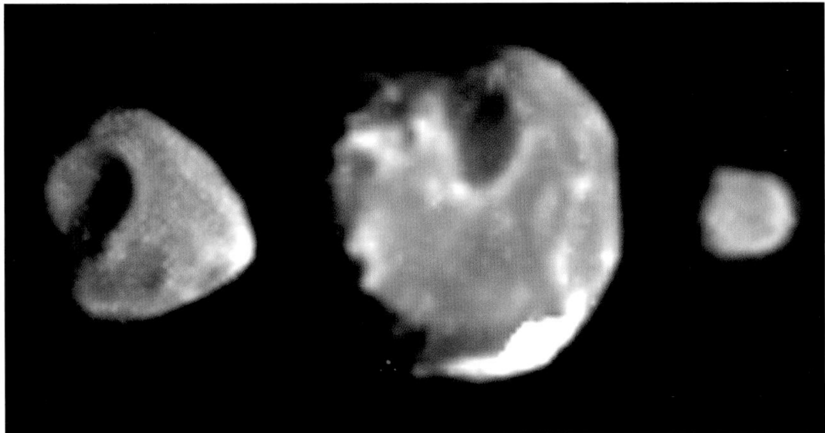

These images of the inner Jovian moons Thebe, Amalthea and Metis (left to right), taken in January 2000 by Galileo, are the highest-resolution images ever obtained of these small, irregularly shaped satellites. The images show surface features as small as 2 km across for Thebe; 2.4 km across for Amalthea, and 3 km (about 1.9 miles) across for Metis.

Scientists don't know very much about any of these moons. Even these size measurements may not be very accurate. They are all irregularly shaped, lacking the mass to pull themselves into a reasonably spherical shape. Galileo has, however, revealed some surface features, including impact craters hills, and valleys. Since Io orbits about 422,000 km from Jupiter, and at this close distance is subjected to extreme tidal flexing from Jupiter's gravity, one would imagine that these even closer satellites would be pulled to pieces. However, because they are so small (the largest, Amalthea, having a diameter that is just one-nineteenth that of Io), they are relatively immune to the effects of tidal forces. However, the two closest moons, Metis and Adrastea, orbit inside what is called the synchronous orbit radius of Jupiter. That is to say, they orbit Jupiter faster than Jupiter rotates on its axis. At this distance the satellites' orbits will eventually decay, and they will fall into the planet. Amalthea is the reddest object in the Solar System, and it appears to give out more heat than it receives from the Sun. This may be because, as it orbits within Jupiter's powerful magnetic field, electric currents are induced in the moon's interior; alternatively, the heat could be from tidal stresses.

Even less is known about the 20 (and counting) outer satellites of Jupiter. Galileo has made no close passes of any of them, and it is assumed that in composition and origin they resemble asteroids (indeed, they are all quite possibly captured asteroids). The eight outer Jovian satellites that have been named are, in order from Jupiter: Leda, Himalia (the largest at 85 km in diameter), Lysithea, Elara, Ananke, Carme, Pasiphae and Sinope. A new moon discovered in 1999, has been given the designation S/1999 J1. The rest of the recently discovered outer moons, designated S/2000 J2 to J11, are all less than 5 km across.

Jupiter's rings

Saturn, as everyone knows, is the ringed planet, its colourful girdle one of the most spectacular features of the Solar System. But all the gas giants are now known to have ring systems. Saturn's main rings are thought to consist of trillions of car-sized ice fragments, maybe leftover material from a "failed" moon, or possibly the remains of a moon torn to pieces by tidal forces. Jupiter's rings are rather different. Invisible from Earth, they are formed from dust kicked up as interplanetary meteoroids smash into the giant planet's four small inner moons. "We now know the source of Jupiter's ring system and how it works", says Cornell astronomer Joseph Burns. "Rings are important dynamical laboratories, [revealing] processes that probably went on billions of years ago when the Solar System was forming from a flattened disc of dust and gas."

The "four stars that wander around Jupiter as does the Moon around the Earth" first seen by Galileo Galilei on a cold winter's night in 1610 have now resolved themselves into a remarkable quartet of worlds. No longer points of light, or even discs, we see them now as living, breathing mini-planets that are among the most complex and dynamic worlds in the Solar System. The Galilean satellites mimic the Solar System as a whole – Io is heavy and iron-rich. Callisto and Ganymede much less so, being made predominantly of ices. There is also a gradient of geological activity: Io is extremely active, Europa somewhat less so; Ganymede hints at a turbulent past, and Callisto looks to be as dead as a doornail (although there are hints even here of interesting things going on beneath that battered crust).

The reason for this pattern must be tied up with Jupiter's formation.

The giant planet condensed from a vast gaseous envelope that was surrounded by a flattened "proto-satellite disc" from which the moons formed. This is analogous to how the planets formed around the infant Sun, and is quite different, say, to how Earth's satellite came into being. The Moon is now thought to have been the result of a colossal impact with a Mars-sized object, which blasted the future Moon-stuff off the Earth's surface and into orbit, where it gradually accreted into a single object. As Washington University planetary scientist William McKinnon says, "So it is by birth as well as by organization that the Galilean satellites, along with Jupiter, can be called a miniature planetary system."

Jupiter continues to shape the evolution of its retinue through the action of tidal heating. Big rocky planets like the Earth and Venus remain hot inside because of the radioactive decay of potassium-40 and thorium in their interiors. Big planets generate a lot of heat, and have thick enough insulating outer layers to trap the heat within. Mars is smaller, and gives the appearance of having "cooled off" substantially in the last billion years (though it is probably still active in some places). The moons of Jupiter are smaller still, and the volcanic activity seen on Io and inferred on Europa cannot be explained by radioactive heating alone. Tidal forces must be largely responsible for this heating. Io receives a colossal amount of heat, and radiates colossal amounts of heat into space, its crust rising and falling by tens of metres as it swings around Jupiter. Europa receives somewhat less heat – any volcanoes beneath that ice are hidden from view, although the great number of curious hillocks on its icy surface are maybe indicative of plumes of hot material rising from the depths. Ganymede is the outermost of the moons that could be subjected to tidal heating, although there is little evidence that receives enough to reshape its surface today. Only Callisto is dormant. A fascinating family of worlds.

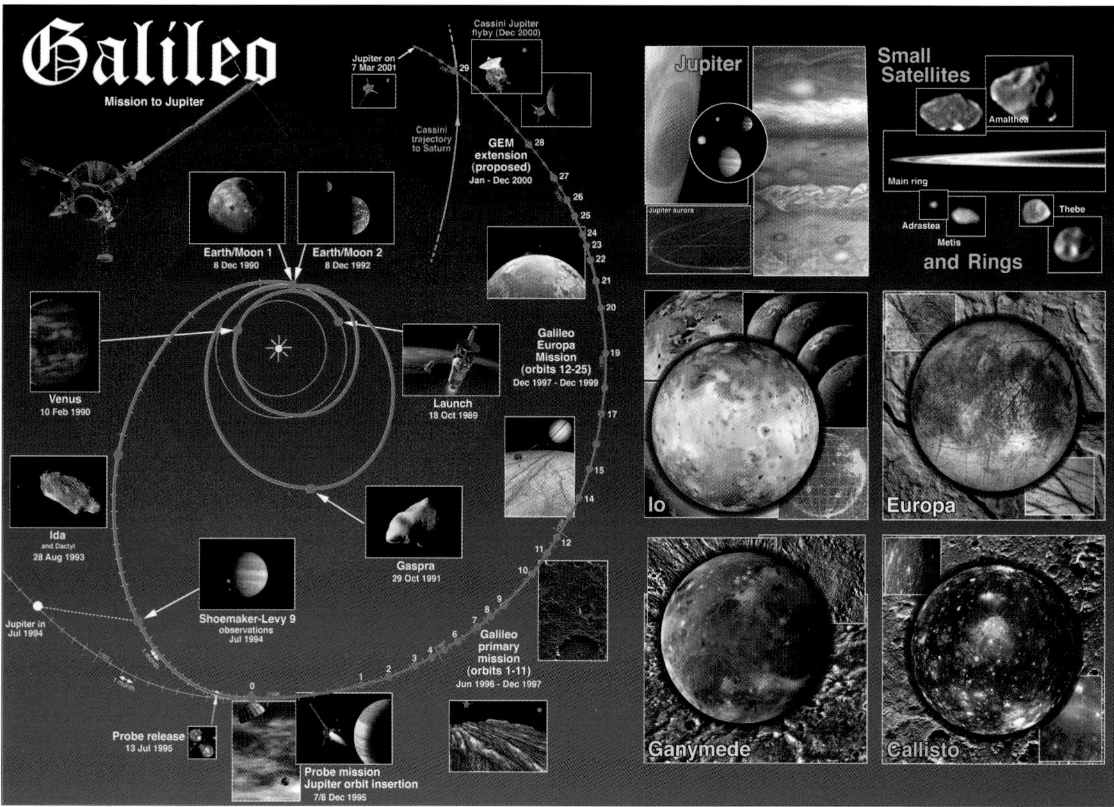

The Galileo mission was a success, by any measure. The spacecraft is now nearing the end of its mission, but scientists will be analysing the results it has sent back to Earth for decades to come.

Chapter eleven

GALILEO – THE END

THE PLANNING FOR GALILEO'S Extended Mission had begun right at the start of the Primary Mission. After arrival in 1995, the mission designers were already hard at work plotting a further orbital tour, concentrating on Europa and Io. As we have seen, that has been a stunning success, producing some of the most unexpected – and in the case of the pictures, beautiful – data from the entire mission. At the time of writing, early in 2001, Galileo was still going strong. This was not entirely expected. The spacecraft was designed with plenty of redundancy, and its systems are extraordinarily robust. But there is always the element of surprise. A stray micrometeorite or a particularly vicious blast of radiation could easily knock out a key component or circuit, ending the mission in a flash. As current project manager Jim Erickson says, "You have no idea how well the spacecraft's going to behave, whether it's going to fall apart, whether the electronics are going to hold together. You don't know, to give an example, if you are still going to have a camera that works. But mostly, you don't know how much funding is going to be available to keep the thing flying."

Funding is perhaps the key component of any space mission. It seems extraordinary that, after spending over $2bn designing, building and launching what is arguably the most impressive robotic space probe in history, there should be a problem with keeping Galileo operating. After all, the cost of the Extended Mission was about $30m – just one-sixtieth of the cost of building the probe. Surely that was money well spent on what would effectively double the returns from Galileo? But NASA, and the Congressional committee that funds it, doesn't always see it that way – witness the cancellations of Apollos 18 to 20, saving a few tens of millions of dollars when NASA had already spent billions getting to the Moon. But, says Erickson, the money was found. "Everything had to come out of it – my salary, the salaries of the team, keeping our investigators funded, getting the data and archiving it. Making sure the calibrations are done. It's all got to come out of that."

With the advent of the Hubble Space Telescope in 1989, astronomers had another way of getting crystal-clear pictures of the outer Solar System. This image of Jupiter and Io taken by Hubble rivals some of the best pictures taken by space probes.

The "Ice and Fire" extended mission is now nearly over. Galileo has a further encounter with Ganymede planned for the end of 2000, and then a first in the history of space exploration. At Christmas 2000, Galileo collaborated with Cassini, JPL's Saturn-bound probe, in a joint observation of Jupiter. The two spacecraft passed by within a few million kilometres of each other, affording an unprecedented opportunity to carry out simultaneous observations of Jupiter from two platforms. Now, a "Millennium Mission" lasting through 2001 and into 2002 has been agreed. Erickson says: "Galileo is going to run out of fuel, in a year or two – certainly by the end of 2002. There is enough fuel to go back to Europa, but now there are some further issues to think about." The most significant of these "further issues" is the possibility of life on Europa, and the slim but distinct possibility that Galileo could wipe it out.

"All these worlds are yours – except Europa. Attempt no landings there."

In *2010: Odyssey Two*, Arthur C. Clarke imagines that humanity will be prevented by a race of godlike aliens from colonizing Europa because that is where intelligent life stands the best chance – unmolested – of emerging in the Jovian system. Now, in a bizarre case of truth imitating fiction, NASA is in the same basket, although the instructions to leave Europa alone have come from the mundane offices of NASA's HQ in Washington DC rather than from a giant extraterrestrial monolith.

It seems to be rather over-egging the pudding to suppose that a little space probe could eliminate the life in Europa's ocean before we have found a single microbe, but NASA is taking the threat very seriously indeed. Galileo was not sterilized before it left Earth. Probes designed to land on planetary surfaces, like the Viking landers, *were* sterilized (though it is now accepted that the level of debugging here was woefully inadequate). The Russians took much flak for allowing their Mars landers to depart for the Red Planet without a proper wash – one even contained a fabric flag signed by Leonid Brezhnev, presumably contaminated by billions of microbes that formerly lived on the hand of the late General Secretary. (Mars got its own back – not a single Soviet or Russian mission to the Red Planet has ever succeeded.)

Scientists are worried about two things when it comes to allowing terrestrial microbes to come into contact with another planet. The first and most likely problem is that somehow our bugs will be able to set up home in their new environment. If later probes triumphantly detect life on that planet, scientists will have to accept that the "extraterrestrials" they found may not be alien at all, just the descendants of those earlier Earthly interlopers. An analogy would be Christopher Columbus running into a bunch of Norsemen in the New World, descendants of Leif Eriksson's party from nearly five centuries before. Presumably DNA analysis would prove the Earthly link, but unwitting cross-contamination is the exobiologist's nightmare. *Natural* cross-contamination equally so, although there is nothing we can do about that. Calculations have shown that both Earth and Mars have flung trillions of tonnes of material at each other over the millennia, blasted off the surfaces of the two planets by asteroid impacts. It is possible that life on Earth is all descended from bugs that made the multi-million-kilometre voyage across the void four billion years ago.

The second possibility is even grimmer. When foreign species are suddenly introduced into long-established ecosystems the results are usually unhappy for the natives. Rats have decimated the fauna of the Galapagos, and rabbits have wreaked havoc in Australia. Off the coast of Baja California, there are islands that have been turned into barren wilderness by the incessant chomping of imported goats. Perhaps humanity's most devastating act has not been the environmental wreckage of habitat destruction, pollution or even global warming, but the mixing of species brought together by train, boat and plane. Already, large areas of the Earth are becoming homogenized as invading species destroy local ecosystems.

The findings from Europa have forced NASA to completely rethink its position on Galileo's demise. Most scientists are agreed that there is a colossal ocean of warm (by Jovian standards) liquid water under the solid ice of Europa's surface. Early calculations indicated that the ocean was at least 50 km down. More recent work based on the magnetometer data show that it is much closer to the surface, maybe only 7 km at the most. No one knows if there is life in this ocean, but there is probably as good a chance of finding aliens in the Europan depths as anywhere. Maybe life thrives on or near the surface, along the ridges in warm slush brought up from the depths. If Galileo crashes into Europa by accident, there is a tiny but finite risk that some surviving terrestrial microbes could thrive there, living off the nutrients in Europa's waters – maybe even living off the *inhabitants* of Europa's waters. Within a millennium, the terrestrial invaders could have annihilated an entire ecosystem. It is an unlikely, even remote scenario – hundreds of millions to one against, perhaps – but one that NASA is taking seriously. It would be very bad press for the agency to discover an alien biosphere only to wipe it out in short order. "Oops, we found ET, but we killed him" would not go down well.

Out of harm's way

"I think it is fair to say that we don't want Galileo hitting Europa", says Jim Erickson. "It was treated in a manner which makes it pretty clean, but clean does not mean sterile, and you would hate there to be any risk whatsoever of contamination." He explains that Galileo's orbit is not stable. The complex dynamics of Jupiter, four large moons and a couple of dozen small ones, not to mention random factors like meteorite strikes, mean that plotting Galileo's movements long into the future becomes a rather imprecise science. "In life, there are no guarantees. If you put something into orbit, you might think that you can guarantee that for the next fifty years it's not going to hit Europa. But then you have to ask, what if you can't get back there in fifty years to check? What if it takes two hundred years? Or a thousand? Can you guarantee that Galileo will not hit Europa in a thousand years? The answer is 'No'."

JPL has several options now. An independent team of experts called the Committee on Planetary and Lunar Exploration (COMPLEX) has been asked to assess each one, taking into account the desire to minimize the risk of hitting Europa and maximize the amount of science that can be extracted from a still-functioning Galileo. It is now looking likely that Galileo's fuel will run out before the electronics fry, and once the fuel has gone the spacecraft will no longer be manoeuvrable. Therefore something has to be done before 2003 at the latest.

One option, explains Erickson, is to try to fly Galileo out of orbit from Jupiter. Galileo has enough fuel to escape from its orbit, but there remains a slight chance that it will one day be captured again. "We can make it escape from Jupiter, but the problem is that we would not be able to control where it went from then on. You haven't eliminated the possibility of it hitting Europa that way, you have just made it smaller."

Theoretically, Galileo could even be set on a course that would bring it back to Earth, though it would be going so fast we could never retrieve it. A novel idea floated was to present Galileo to a team of students who would fly the ailing spacecraft until it irretrievably broke down. But that option would not have protected Europa. The spacecraft contains no self-destruct mechanism, so that only leaves NASA the option of deliberately crashing Galileo into something to which it can't possibly do any damage. That means Jupiter itself. Since Jupiter's boiling atmosphere is unlikely to support life, NASA's planetary protection team has decreed that the giant planet itself will become Galileo's graveyard. "We are going to do what headquarters tells us to do here", Erickson says. "Bear in mind that without the original science done by Galileo you wouldn't have known there was an issue with Europa, so that shows that you don't get a return unless you have some risk, but there does come a time when you have to weigh those risks up." It seems that the orbiter is destined to share the fate of the atmospheric probe, and become at one with the atmosphere of Jupiter.

The current plan is to keep Galileo flying through 2001 on a so-called "Millennium Mission", which will most likely concentrate on Io. It is ironic that the world snubbed by Galileo on its arrival at Jupiter will now provide the backdrop for its final days. Erickson says, "Io has got to be the place that will give the most return for a couple more encounters. We didn't get the opportunity to see much of Io during the Primary Mission, but now we have gone back it looks great, somewhere we want to learn more about."

Galileo will be sent into a series of incredibly close encounters with Io, down to an altitude of maybe 100 km or less. The close passes of Io during the Extended Mission allowed Galileo's camera to resolve objects down to about 7 m. "That's not fine enough to see this table," says Erickson, pointing at his JPL-issue desk, "but this office will show up". Any closer would be pointless. At, say, 50 km Galileo's camera could resolve objects as small as this book, but at that altitude the ground would be rushing underneath so fast that any image would be blurred.

One place Galileo won't be going back to is Europa, even though the scientific payback of a series of super-close flybys could be tremendous. "Common sense says that this is not an option: after all, this the one place we are trying to avoid," Erickson points out. "But that isn't to say that there aren't people who would really like to do that. So you say to them, 'Well, will going in really low solve the issue once and for all whether there is liquid water there?' And they say, 'It might tell you something, and it might not. And it isn't going to deal with the issue of life because you can't do a landing and look for it.'"

So, sometime in 2002 or 2003, a final set of commands will be sent to Galileo. Its motors will fire for the last time, and using the gravities of Io and possibly Europa to send it on its way, it will be directed to its doom. Lacking the sturdy protection of the atmospheric probe, it will break up quickly as it slams into Jupiter's atmosphere, its constituent parts glowing red- then white-hot before finally evaporating. A dramatic end, and perhaps a sad one for such a successful machine. But there is clearly no alternative. The risks of letting Galileo live – perhaps as some sort of relic to be collected centuries hence and returned to the Smithsonian Institution – are too great. In any case, Galileo will turn out to have been a reconnaissance mission. In the pipeline are plans for a dramatic return to Jupiter and a concerted assault on the mysteries of Europa. By the end of this century, humanity may be treating the worlds of Galileo – all of them – as its own.

Galileo's legacy, NASA's conundrum

Depending on who you talk to, Galileo and Cassini – its all-singing, all-dancing successor now on its way to explore Saturn – are either the "last of the Rolls Royce missions" or the "last of the dinosaurs". Both cost billions of dollars, and have soaked up the expertise of hundreds of scientists, engineers, planners and administrators. People such as John Casani, Torrence Johnson and Margaret Kivelson have devoted much of their working lives to these missions. Galileo, like the Voyagers before it, was born of a get-it-right attitude that had prevailed at NASA since the days of Apollo. But even by the time of Galileo, a wind of change was blowing through the corridors of NASA, and in JPL in particular.

The Space Shuttle programme was eating up a huge chunk of NASA's resources, and the science missions suffered. There was no chance of building a pair of Galileos, for instance – something everyone at JPL seems to agree would have saved a huge amount of grief (it is highly unlikely that *both* antennas would have failed). Mission after mission was cancelled or curtailed. The principle of doubling up was abandoned – there was one Magellan, one Ulysses, one Galileo, one Cassini. When Dan Goldin took over NASA in 1992, almost his first act was to institute a policy of Faster, Cheaper, Better. This meant developing a series of missions all on a shoestring, the idea being more missions for your money. Redundancy was abandoned – "FCB" missions like Mars Pathfinder had just one of everything. Mars Pathfinder – which was an outstanding success – was hailed as a triumph for the new NASA philosophy. Unfortunately this triumph was short-lived. Two subsequent Mars missions – which, like Pathfinder, cost a couple of hundred million dollars rather than billions – ended in inglorious failure, one crashing because of a farcical confusion over metric and Imperial measurements, the other because its landing system failed. Inquiries were launched and the finger of blame was pointed, but JPL old-timers say the real reason was the new atmosphere of cost-cutting.

John Casani, the man who got Galileo off the drawing board and led the inquiry into the 1999 Mars losses, says the new philosophy has shown its limitations. "We clearly went too far with Faster, Cheaper, Better on these two Mars missions. The mistakes that were made were very human mistakes, which were avoidable. Having been made, they should have been caught. They should have been uncovered and corrected before launch. And the reason that they weren't uncovered, and maybe to a degree the reason they were made in the first place, is because we tried to cut back the cost of the mission. There are, after all, only a couple of ways you can attack the cost. You have to build the hardware, you have to build the science instruments, you have to put it all together, you have to rate the software, you have to test that and you have to launch it. You don't have to do all the quality inspection or the double checking and oversight and review. And you don't have to have all the safety nets. You know, you can say, 'Hey, we've got people here who've done this before. They're good enough.' But we went too far. We were trying to improve productivity, and reduce the cost of the workforce, staffing costs. When you do that you expose yourself to mistakes of a very human kind."

There are dark mutterings in the corridors of JPL that the success of the Pathfinder mission, in 1996, was one of the worst things that happened to NASA. If the little Mars probe and its rover had smashed themselves to bits on a rock, it would have called into question the whole FCB philosophy early on, before it was allowed to soak up more resources and do more damage. Casani admits as much: "I wouldn't call it a bad thing, but I think there were some wrong lessons learnt from the Pathfinder mission", he says.

Eventually, if space exploration is to continue, then a compromise will have to be reached. Casani says, "My opinion, and I think this is shared by many people, is that we don't need to go back to the Rolls Royce era, you might call it, of Cassini and Galileo. I think most people believe that the more rapid missions, with faster turnaround, really is a healthier way to go. To me, Faster Better Cheaper means reducing the cost of success. It doesn't mean taking more risks, although Dan Goldin will say, 'If you're not losing one or two out of ten then you're not pressing the envelope hard enough. I expect you to lose one.' Well, that's fine coming from him, but as a project manager I could never accept the fact that I was going to put something on the launchpad that was going to be the one out of ten. So it strikes me that when you talk about what's an acceptable failure rate, you're talking about it in the larger sense. It's like saying that in the US we probably have two to three hundred automobile fatalities every day. And we accept that. But it's not acceptable for you or me or anybody to put their family in the car and go to the beach or market or church and die. I mean, when you're in the car you don't expect to be one of those two hundred people – that's not acceptable. So if you're the driver, if you're the project manager of this car, it's not OK to fail and you're going to do everything within the limitations of your resources to ensure it doesn't, to manage the risks and to put in an appropriate level of checks. You have to accept that people are human and humans make mistakes. But you don't go in saying, 'Oh if I lose this one it doesn't matter because one out of ten or two out of ten is an acceptable failure rate.' It's OK for Dan Goldin to say that he'll accept one or two failures out of ten but it's not OK for us to accept that."

Neal Ausman, now retired, is more outspoken. To him, Faster, Better, Cheaper is both wasteful and encourages a sloppy attitude. "Is it good to have an attitude that says, 'It's inexpensive so if it fails it's not a big deal'? I think for engineers that is a dreadful, dreadful approach. Engineers need to work to find the right answer, that's the way: engineers don't deal with nearlys, they deal with 'here's the equation, and here's the answer'. It takes a lot of experience to fly complicated missions like Galileo. One of the things that happens when you have a lot of smaller but still very complicated missions is that you do not create a training ground to produce the seasoned veterans you need to steer the ship. You get a lot of very bright, a lot of very talented people. But 'bright' doesn't always answer every question, and a little bit of the experience can be more useful than extraordinary brains." Galileo, Ausman concludes, "was a spacecraft that we felt could not be allowed to fail."

The end of Galileo does not mark the end of humanity's efforts to explore the Solar System. An equally large, equally expensive probe, Cassini, is on its way to Saturn. Follow-up missions to Europa and Io are planned. Cost, as always, will be the key determiner. But if NASA – and Congress – hold their nerve, the next three decades could be another golden age of space exploration.

Cassini took this extraordinary picture of Jupiter, showing the Great Red Spot, in December 2000, on its way to Saturn.

Chapter twelve

THE FUTURE: AFTER GALILEO

GALILEO – TROUBLED FROM THE start, and plagued by delays, disasters, cost-cutting and technical hitches – has revolutionized our knowledge of the Jovian system. It has detected an ocean on Europa (and perhaps on Callisto, and Ganymede too), discovered hundreds of volcanoes on Io and found the engine that drives Jupiter's weather. Impressive as that may be, Galileo has only scraped the surface of what there is to be learned about these extraordinary worlds.

Scientists debate which is the most impressive of Galileo's discoveries – some at JPL say it is the extent and source of Io's volcanism, while others cite the completely-by-chance discovery of Dactyl, the tiny moon orbiting the asteroid Ida. But the majority of scientists who have worked on the mission agree that near the top of any Galileo top ten must surely be the discovery of an ocean on Europa. This is also the part of the mission that has grabbed the public's attention the most. The pictures of the Conamara "ice rafts" released by JPL made the front pages of newspapers all around the world. It looked as though Arthur C. Clarke's imagined sub-ice ocean was real. What seized the public's interest, of course, was the possibility of life on Europa, which just a couple of decades before had been regarded as one of the most featureless places in the Solar System. Suddenly, little Europa had been catapulted above Mars as the most likely home for ET, and NASA wasn't going to miss the opportunity to lobby for a follow-up mission to this intriguing world.

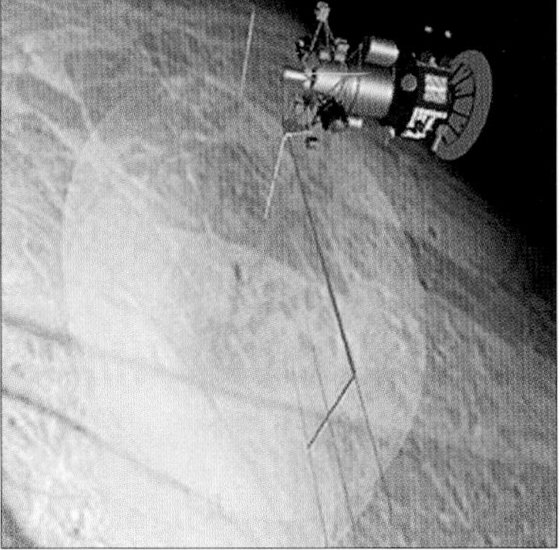

By 2010, JPL hopes to have launched the Europa orbiter, which would be the first manmade object to orbit the moon of another planet. Using radar, it would measure the thickness of the Europan ice and show once and for all whether an ocean lurks underneath.

The history of NASA's attitude to extraterrestrial life is long and complex. In its early days, the Agency was almost fanatical about keeping a lot of clear blue water between itself and the "UFO brigade". The flying saucer boom started just a decade before NASA's creation, when in June 1947 Kenneth Arnold saw "skipping saucers" flying through the air over Mount Rainier, Washington State. Also that year was the infamous Roswell incident, in which a mysterious object, almost certainly a military weather balloon, crash-landed in the New Mexico desert. The result was decades of speculation that aliens had landed and been captured by the military, or that the military had access to hardware somewhat more advanced than it was owning up to. Roswell spawned a whole, quasi-religious nonsense of alien abductions, visitations by the mysterious "grays", the whole ET shebang and *The X-Files*.

No wonder NASA wanted to distance itself from all that. Its stated mission was simple: to explore space. Its less-often stated mission – to first catch up with, then beat, the Soviets – was equally straightforward. Searching for aliens was not something of interest to the Agency – it didn't even start employing biologists to begin a serious search for alien life until the 1990s. NASA was all about hard science and engineering, not science fiction. When the Agency took over the military rocket facility outside Pasadena in the late 1950s, the new name chosen for it, "Jet Propulsion Laboratories", was a deliberate misnomer. "It should have been called Rocket Propulsion Labs, but in the 1950s they thought that sounded silly", says JPL old hand Norm Haynes.

In any case, finding extraterrestrial life looked to be a hopeless task. The Moon, it was quickly established, was dead. Venus (too hot) and Mercury (ditto) were hopeless, so that left Mars. When the Viking landers found no sign of life on the Red Planet in 1976 that pretty much sealed it for NASA: looking for aliens simply became something no self-respecting scientist would do. The Agency's involvement with the SETI project – the Search for Extraterrestrial Intelligence, which used a network of radio telescopes to listen out for signals from alien civilizations – was quietly shelved. (SETI survives, indeed thrives, thanks partly to funding from private sources and universities, and mostly to the seemingly limitless energies of Jill Tarter, its scientific director, and its late standard-bearer, Carl Sagan.)

The trouble was, although NASA's chiefs may not have been very interested in ET, the public certainly was. The Agency depends on public interest for its survival: if it can't make the front pages with daring missions and amazing discoveries, support on the street wanes, and the purse strings are quickly tightened by Congress. Public interest remained high in the 1960s thanks to Apollo. But then pictures of spacesuited men bouncing around the dusty lunar lava fields became commonplace, and NASA shelved the project. "It's amazing, they actually got bored with exploring another planet", said Apollo historian David Harland. The fact that Apollo, having done its job of beating the Communists to the Moon, was starting to reap real scientific rewards (and yet more science was planned for the shelved Apollos 18, 19 and 20 than for the first seven missions combined) seemed to count for nothing. The mighty Saturn V rockets built to launch these ambitious missions now rust slowly in the grounds of Cape Canaveral, a sad testament to short-sighted cost-cutting and to what might have been. NASA then looked to the Space Shuttle as its saviour. The troubled spaceplane had the singular advantage of needing human pilots, and NASA reasoned (probably quite correctly) that if it had astronauts it could maintain a modicum of public interest.

It is very odd that it took NASA so long to hit upon the idea of using extraterrestrials to boost its claim for precious public money. The first sign that the Agency had rethought its attitude to ET came in August 1996, when it announced to a stunned world that it had found evidence of fossilized bacteria-like structures in a lump of rock thought to have been blasted off the surface of Mars. The meteorite, designated ALH 84001, was picked up in the Allen Hills (hence the "ALH") region of Antarctica in 1980 and determined to be of Martian origin because minute gas bubbles trapped inside it exactly matched the atmosphere of Mars sniffed by the Viking landers. NASA claimed that microscopic tubular structures found inside the meteorite could, just could, be the remains of ancient organisms that had presumably evolved on the Martian surface. The announcement was brilliantly timed – August is a very slow month for newspapers, and the media seized upon the story as if it were manna from heaven. President Bill Clinton declared himself interested, and a beaming NASA seemed to be saying, "Well, you thought space was boring, but look at this! Aliens! OK, it's not little green men, it's bugs, and dead bugs at that, but you never know what might still be lurking out there. I think we'd better go take another look."

In fact, announcing the "discovery" of Martian life turned out to be rather premature. When scientists went back and took a closer look at ALH 84001, other explanations for the tubular "bacteria" began to seem more likely. Terrestrial contamination could never be ruled out (after all, the rock had sat on the Antarctic ice for thousands of years), and mineralogists pointed out that the structures could easily have been created by non-biological processes. In other words, no bugs, just interesting crystals. At the time of writing, new evidence has emerged that ALH84001 does indeed contain ancient fossils. This one will run and run.

Spurred on by the possibility of life on the Red Planet, NASA opened a new Astrobiology Institute to examine the possibility of life on other worlds. Its Martian finding – and attempts to secure money for missions that might uncover extraterrestrials – was boosted by discoveries that terrestrial life can exist in extreme environments, for example clustered around hot water "black smokers" deep on the ocean floor. These extraordinary phenomena, discovered only in the 1980s, are found where superheated, mineral-rich waters are ejected into the near-freezing abyssal ocean. Like inky submarine geysers, smokers act as a source of chemical energy for whole ecosystems, starved of sunlight but deriving their nourishment instead from sulphur compounds and other exotic chemical species. Life may in fact thrive in an even more unlikely environment – 3 km beneath the ocean bed. Core samples drilled from the seafloor in the Indian Ocean west of Australia have been found to contain odd, tubular structures that look remarkably like the Martian "bacteria". British geologist Phillipa Uwins claims to have seen these things growing in the lab, and has named then nanobes, because their world is on the scale of the nanometre – one billionth of a metre, far smaller than any known cellular organism, and smaller even than most viruses.

The "Cryobot": one day, scientists hope to build a machine like this to tunnel beneath the ice of Europa and explore the ocean underneath.

If Earthly bugs can survive around black smokers and 3 km below the seafloor, then alien microorganisms might well be able to survive in all manner of places. Dick Taylor of the British Interplanetary Society, a former physicist who has turned his attentions to the study of possible alien life environments, thinks that life may well thrive not only on (or rather *in*) Mars, but also under the lunar surface as well. The discovery of the black smoker ecosystems obviously has the most pertinence to Europa. If Europa has a liquid ocean, then there might be volcanoes on the seafloor, just as on Earth. One view is that Europa is basically Io, a hellish cauldron of volcanoes, but covered with water. When Galileo dies, it will have left the question of Europan life unanswered. As to the reality of the ocean, very few scientists are willing to stick their necks out and say the data does *not* point to liquid water under the surface. But while it is persuasive, the magnetic and photographic evidence gathered by Galileo that points to a Europan ocean is not 100 per cent conclusive. There *could* be ice all the way down. So the people at JPL, and the world's growing community of astrobiologists, would very much like to go back to Europa, this time with a dedicated spacecraft that would spend months or years in orbit, equipped with radar probes capable of seeing straight through the ice and probing the ocean depths.

The design for a Europa orbiter is already fairly well advanced. The spacecraft itself need not be too complicated, containing tried-and-trusted technologies like solid-state cameras, magnetometers and, vitally, a radar instrument that will prove the existence of liquid water once and for all, and tell scientists exactly how deep the ice crust is. Studying Europa from close up is not difficult – Galileo has already done that superbly – but getting into orbit around a moon of Jupiter is another matter again. Galileo needed a large engine to slow itself down enough from its long, high-speed interplanetary odyssey, so that it could fall into the gravity well of Jupiter. But to go into orbit around a world smaller, lighter and with less gravity than our Moon requires a lot more braking power from a much larger engine.

The Europa orbiter being designed at Pasadena is in fact mostly engine. A giant version of the motor that brought Galileo into its Jupiter orbit dominates the tiny spacecraft, which will be a fraction of the size of the last machine to enter the Jovian system. The plan was for the Europa orbiter – a mission costing hundreds of millions rather than billions of dollars – to be launched sometime in 2003, although the recent failures of two JPL Mars missions have threatened to delay launch by as much as seven years, and the Europa Orbiter team, just like their Galileo predecessors two decades ago, still have no firm launch date. If and when Europa Orbiter gets off the ground, it will hopefully take a fast route to Jupiter, arriving in just three years. Firing its huge engine several times to allow itself to be lassooed first by Jupiter and then by Europa itself, the little spacecraft will then settle into an orbit just 200 km above the icy plains and cliffs of the surface.

With its state-of-the-art cameras, Europa Orbiter will give scientists unprecedented views. Where Galileo could get down to a resolution of 6 m, Europa Orbiter, flying lower and with better cameras (and a fully functioning high-gain antenna), should be able to see objects a metre across or less. If icebergs are moving across Europa's surface, this spaceship should spot them. If cracks appear in the ice, and are filled in with boiling water from below, the probe should spot that too. But it is the Europa Orbiter's radar sounders that will prove the existence of an ocean once and for all. They will probe the ice using radio waves, and the pattern of reflections will give scientists a detailed cross-section through the outer layers of the planet. If an ocean is there, Europa Orbiter will find it.

What happens next very much depends on what Europa Orbiter finds. If there is no ocean on Europa – an unlikely but possible scenario – then interest in this moon will undoubtedly wane, just as interest in Mars waned after the Mariner probes sent pictures back of a cratered world that looked pretty much like our Moon. NASA will then switch its attentions elsewhere. But there are plans to send a whole armada of spacecraft out to explore this world if a Europan ocean is confirmed. Those on the drawing board include the Ice Clipper, which will involve an orbiting probe flying through plumes of ice kicked up by missiles fired into the surface, and performing chemical analysis. If the ocean is teeming with life, then traces of biological compounds such as nucleic acids and amino acids – perhaps even whole, frozen organisms – ought to have made their way to the surface.

By far the most ambitious plan is to send a robot submarine – a "cryobot" – to drill through the Europan ice and plop into the ocean below. According to Seattle oceanographer John Delaney, who has conducted feasibility studies on such a cryobot mission, detecting life in the Europan ocean should be easy. "If it's there, the chemical signal should be pretty unmistakable. You wouldn't need much in the way of sophisticated instrumentation." But keeping in touch with an ice borer as it made its way down through maybe 10 km of ice would be problematic. One solution proposed by JPL scientists is for the drilling machine to drop off a series of UHF radio transponders as it made its way through the crust, building up a miniature communications relay frozen into the ice. "Trouble is," Europa veteran Bob Pappalardo says, "the best way of getting through that ice is to use a radioactive generator, and they aren't too popular at the moment. So we may have to find another way."

Many scientists believe that drilling through the ice is only half the problem of exploring the Europan ocean. By far a bigger source of grief, they say, will be the need to sterilize such a probe before it even hits the surface. So far, human-built machines have landed on three extraterrestrial bodies: the Moon: safely assumed to be sterile, at least in its upper crust; Mars, a place where any Earthly bugs on the surface would soon be fried in the fierce ultraviolet glare; and Venus, where they would be roasted in milliseconds. Europa, however, is different. If there is liquid water there, sloshing around at or near its freezing point, it would be all too easy for a hardy Earthly bacterium to spread like wildfire through the water. Who knows what would happen then? There is a chance that the alien invader would wreak havoc in the Europan ecosystem.

Despite the undoubted problems, no one you speak to at JPL has any doubts that humanity will soon revisit Europa, the "banned" world of Arthur C. Clarke's *2010: Odyssey Two*. According to Torrence Johnson, "We'll land something on Europa within the next ten to twenty years because it is not actually that difficult a job. It is a lot more difficult than landing on Mars, but we are studying all sorts of alternative landing systems. If somebody really wanted to land on Europa, we could launch in a few years because we would use standard rocket technology." Any planetary scientist would give their right arm to explore Europa close up. "You'd find a lot of ice – probably kilometres of it before you got to water, but I think you would get water down underneath that. We simply don't know how deep – that's one of the reasons why we're trying to figure out what we should take on a Europa Orbiter mission."

Europa on Earth

Unfortunately, getting to the Europan ocean may well prove to be beyond our technology for decades (unless we strike lucky and find biological material on the surface). It would be nice if we could find somewhere a little closer to home to study, an ocean-under-the-ice here on Earth on which scientists could hone their drilling techniques, and work to design a craft to explore the Europan depths.

Happily, there is such a place – as remote as anywhere on Earth, but still a damn sight easier to get to than Jupiter. Nearly 4 km under the ice on which sits the Russian Antarctic base of Vostok lies a huge body of liquid water the size of Lake Ontario. Vostok is the most inhospitable place on Earth, famous for its temperatures, which can nudge minus 90°C – and that is before taking the hurricane-force wind chill into account. Lake Vostok, detected in the 1970s by aircraft-born sounding radar (exactly the same technique that will one day be used to confirm or deny the existence of the Europan ocean), is Earth's largest underground lake. Cut off from the outside world for millions of years, this is the most pristine body of water on the planet. And now NASA scientists are proposing to drill through the ice above and explore Lake Vostok, in what is being seen as a dry (or rather wet) run for the exploration of Europa.

Getting to Vostok won't be easy. For a start, there is the awful weather to contend with. The top of the ice is 3,700 m above sea level; the radar soundings suggest that the lake surface itself is just below sea level. Judging from the contours of the surrounding bedrock, the lake seems to be sitting in a rift valley, a place where adjacent crustal plates are being pulled apart. The water is liquid either because of geothermal heating – the lake could lie over a mini-hotspot, a mantle plume similar to the feature that generates the volcanoes of Hawaii, or simply because the insulating effects of all that ice is enough to trap the trickle of heat that leaks out from our planet's interior at every point on the surface. In fact, there are several lakes under the Antarctic ice; Vostok is simply (by far) the largest.

The ringed planet: This exquisite picture of the planet Saturn was one of the highlights of the Voyager missions.

In many ways, the similarities between Europa and Vostok are remarkable. The depth of ice is at least broadly similar. In 1996, Frank Carsey and Joan Horvath, both scientists at JPL, proposed a mission to drill through the Antarctic ice sheet into Lake Vostok. They suggested that a similar mission, using more or less identical technology, could be mounted to Europa at a later date.

The ringed planet: Passing on the torch

Our interest in the outer Solar System does not stop at Jupiter and its moons, fascinating though they are. Galileo has opened up a mini solar system of amazing diversity, and its results and data will keep scientists busy, and arguing, for decades. But a probe even more sophisticated than Galileo is already on its way to Saturn, the next great target in the outer reaches of the Sun's empire. The Cassini probe, built in the same assembly rooms in Pasadena as Galileo and launched amid some controversy on 15 October 1997, is due to reach the ringed planet in January 2004. Cassini is very much a joint mission – between old hands NASA and new kids on the block (at least when it comes to exploring the outer Solar System) the European Space Agency (ESA). Perhaps the biggest problem faced by the Cassini team was a blast of objections by environmentalists, who almost got the launch cancelled. Their fears were the same as those of the well-meaning Greens who tried to scupper Galileo, namely that Cassini's radioactive batteries could contaminate the Earth's atmosphere should the spacecraft explode.

But the launch went ahead, and the flight to the outer planets began flawlessly. At the end of 2000, Cassini swung past Jupiter and, the two spacecraft made the first-ever double observation of another world by deep-space probes. Cassini's camera has already taken some stunning pictures of Jupiter; while Galileo continues to explore this planet, Cassini is rushing onward to its date with the most beautiful object in the Solar System – Saturn.

Saturn, like Jupiter, is a gas giant, a huge ball of hydrogen with no solid surface and heated from within by gravitational compression, heat being brought to the surface by convection currents. Although huge (120,000 km across), it is only a third of the mass of Jupiter, and considerably less dense. (It is one of the clichés of introductory astronomy texts that, if one could find a bowl big enough and fill it with water, and wrap Saturn in highly efficient heat insulation, then the planet would float in it.) Winds blow at high speeds in Saturn – even faster than the Jovian hurricanes. Near the equator, the two Voyagers measured winds about 1,800 kph. The wind blows mostly in an easterly direction. The strongest winds are found near the equator, and windspeed falls off

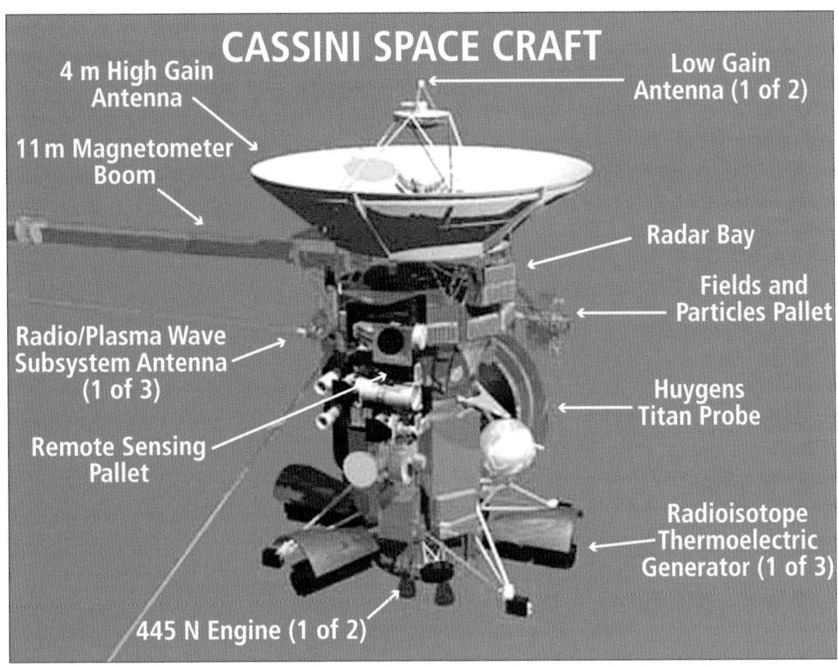

Cassini – the largest and most sophisticated robot spaceprobe yet to be built by JPL.

steadily towards higher latitudes. For years astronomers puzzled why Saturn should have such strong winds – after all, the planet is smaller and cooler than Jupiter, and should therefore have less energy to drive its weather systems. Neptune, at the chilly outer marches of the Solar System, has even faster winds, yet receives almost no energy from the Sun at all and is so small its internal fires must be very dim indeed.

The answer seems to be that, over a certain limit, pumping more energy into a weather system acts to break up fluid flow in the atmosphere, shearing planet-girdling jet streams into a series of swirling eddies and maelstroms. This

A cutaway drawing of one of the radioactive electricity generators on board Cassini that invoked the wrath of the green lobby is why Jupiter's clouds are fragmented into a series of storms like the Great Red Spot, whereas Saturn's clouds are formed into neat bands that are rarely swirled into huge depressions and anticyclones. On Saturn, at latitudes greater than 35°, winds alternate east and west as the latitude increases. The marked dominance of eastward jet streams indicates that winds are not confined to the topmost cloud layer, but must extend downward at least 2,000 km. Furthermore, measurements by Voyager 2 showed a striking north–south symmetry that leads some planetary scientists to suggest the winds may extend from north to south through the interior of the planet.

Cassini, named after the seventeenth-century Italian-French astronomer Giovanni Domenico Cassini, was the last of the Rolls Royce missions. It got the Congressional go-ahead in October 1989, the same month that Galileo finally began its tortuous odyssey to Jupiter. What was then envisaged was CRAF–Cassini, a highly ambitious double mission: CRAF, the Comet Rendezvous and Asteroid Flyby, would explore selected small worlds while Cassini made for Saturn. Between 1989 and 1992 the mission was scaled down, and what was launched in October 1997 was a somewhat "descoped" but still hugely impressive spacecraft, a generation in advance of its Jovian predecessor.

Cassini's long tour of the Solar System on its way to Saturn

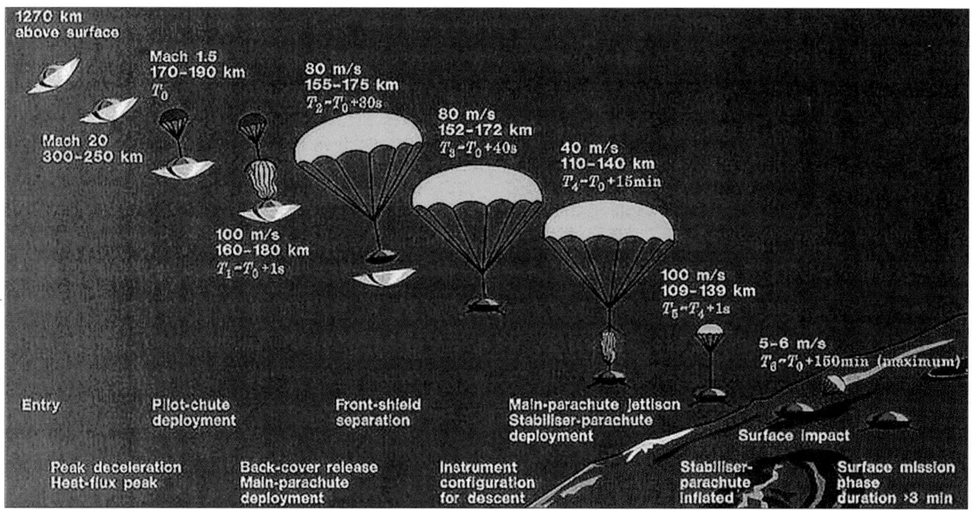

Diagram showing the descent of the Huygens probe through Titan's atmosphere.

Cassini was built with 1990s technology, and its relatively short gestation time means that its equipment is still pretty much up-to-date. Indeed, in its electronics and computing capability it is aeons ahead of what is on board Galileo. Cassini has no tape recorder to stick, tangle or snap. Its Italian radio antenna is a rigid dish, so that with nothing to unfurl there is nothing to jam. Since Cassini was launched by a Titan-Centaur rocket (Galileo's launch vehicle of choice), rather than from a Space Shuttle, its 4m main antenna could be carried aloft as a single unbreakable unit manufactured in the open position. Cassini's computers make those on Galileo look like something out of the Stone Age. Its banks of memory chips not only make a tape recorder redundant (images and data are stored as solid-state memory, so there are no moving parts), but also enable Cassini to perform much more delicate manoeuvres, and to respond faster and more accurately, to real-time commands from Mission Control.

Although Cassini did not have to be launched from a Space Shuttle, it is still having a hard time getting out to the Saturnian system. Just like Galileo, it has undertaken a grand tour of both the inner and the outer Solar System, using gravitational slingshots courtesy of Venus, Earth and Jupiter to propel it to its target. From Earth, Cassini headed towards Venus; after being boosted by Venus it looped around the Sun and came back for a second Venus encounter; from Venus Cassini headed directly to Earth; and from there it headed for Jupiter.

The primary design of Cassini was done at JPL, mostly by Galileo veterans. The initial project manager was Galileo chief John Casani, and now in charge is former Galileo Project Manager Bob Mitchell. He says that Saturn is the next great goal for exploration – another suite of worlds waiting to be unveiled. "Saturn represents a new, unexplored environment for us, unexplored in the sense that we haven't yet got up real close and done much *in situ* investigation. We've been there with the Pioneer spacecraft, and with the two Voyagers, but they were flybys. The next step in the exploration of the planets after flybys is orbiters. and then landers."

Cassini is due to make sixty orbits around Saturn, exploring the planet from a distance of just 180,000 km, and scrutinizing its complex braided ring network and its litany of moons, getting close-up views of at least eight of them. That is just the primary mission; an extended mission, as with Galileo, could double the science return from the $3bn investment.

Like Galileo, Cassini is not one but two probes. The "mother ship" orbiter will despatch a small sub-probe, not to dive into Saturn's atmosphere but to land on the surface of Saturn's mysterious moon, Titan. Cassini will release the Huygens probe, built by the European Space Agency, as it swings by Titan in November 2004, and the sub-probe will parachute down on to the moon's surface. Huygens is named after the seventeenth-century Dutch astronomer Christiaan Huygens, who in 1655 discovered Titan, and also established the true shape of Saturn's rings.

Titan, along with Europa, is a world that thousands of scientists would like to take a closer look at. For a moon, Titan is huge – second only in size to Ganymede, and bigger than Mercury. Uniquely for a satellite it has a thick atmosphere, in fact 60 per cent denser than Earth's: the gas giants aside, this is the second thickest air blanket of any planetary body known, after Venus. In composition its atmosphere is quite similar to Earth's, even today – a familiar 80 per cent nitrogen. But even though Titan has similarities to Earth it also has differences, and the most obvious is the cold. The thick air provides some warming effect, but this far from the Sun the surface of Titan stands at a chilly minus 180°C, far too cold for any chemistry based on liquid water. The remaining 20 per cent of the atmosphere is a noxious mixture of

This artist's impression shows the European lander Huygens parachuting down to the surface of Saturn's enigmatic moon Titan. To the left can be seen towering cliffs made of water ice, to the right an ocean of liquid hydrocarbons. The view of Saturn in the sky is artistic licence – Titan's atmosphere is probably too murky to see through into space.

ammonia and various hydrocarbons – butane and methane, all combining to form dense orange organic smog that shields the surface of Titan from view. When Voyager 1 swung past in 1981, it took a picture of a dull, orange ball – the most boring photograph in its album, JPL admitted. But what lies beneath the smog may be far from boring: scientists think that today Titan is probably very similar to Earth about three or four billion years ago, but deep-frozen.

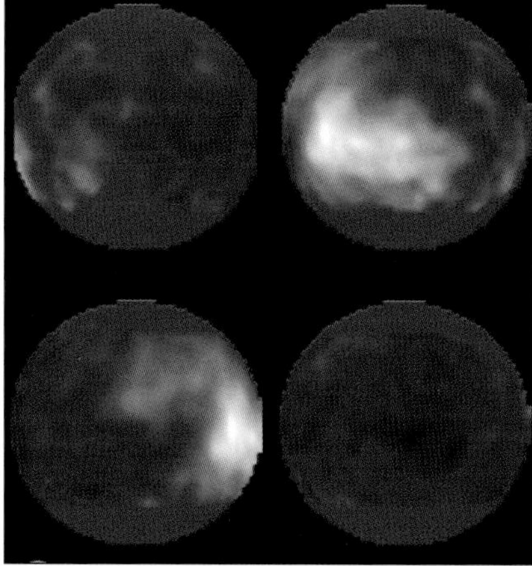

Titan is shrouded in clouds, and its surface is hiden from view. However, the Hubble Space Telescope is able to peer through the clouds in the infra red, and has revealed distinct bright and dark areas. Scientists believe that the bright areas could be the tops of mountains, or continents, probably made of pure water ice, washed clean by methane rain. The dark areas could represent huge seas of liquid methane and propane.

Titan was and remains a mystery, but like Venus the prying eyes of modern astronomy are slowly lifting their veils. The Hubble Space Telescope and land-based observatories have used their infrared capabilities to peek beneath the clouds of Titan to the enigmatic surface, revealing "bright" spots that may be continents, and dark spots that may be oceans of liquid methane or ethane. There is a danger that we are repeating the mistakes of previous generations who saw "seas" on the Moon, and that what we are looking at are no more than dry and dusty lava fields that have never seen a drop of water in their long existence. Nevertheless, there is a growing consensus that Titan could be the one place in the Solar System apart from the Earth with large area of open sea or ocean – albeit of inky black liquid hydrocarbons rather than water. If this is true, then the landscape of this world must be a truly bizarre place. Continents of rock and ices rising majestically from great oceans of liquid ethane and methane; towering "icebergs" of frozen propane; 50 m waves lashing at the frigid shores; pure water-ice mountain ranges, hard as granite in the frigid temperatures, lashed and eroded by methane rain, rivers of ochre-brown hydrocarbons cascading down their flanks. Above, a dull pink haze, perhaps with icy squalls, blizzards of frozen propane snow. And rain. Rain like nothing on Earth. Drops of liquid methane the size of tennis balls float down gently, at the speed of a descending snowflake here on Earth, Titan's weak gravity allowing them a leisurely descent through the thick atmosphere.

Every now and then, perhaps, comes a break in the clouds – allowing a glimpse of Saturn, its majestic rings glinting in the faint sunlight. Below, the ground is slushy: dirty-red propane snow mixed with a cocktail of oily aromatic compounds. If an astronaut visited Titan and took a sniff of the air he would reel as his senses were assaulted by the pungent aroma of a petrochemical refinery. A more dramatic vista it is hard to imagine, and with a bit of luck a vista that should be coming to a computer screen near you in November 2004, when the Huygens lander detaches from the Cassini probe and makes its way to the surface of Titan, sending back pictures until its batteries give out, perhaps an hour after landing.

Down to Titan

After the probe is released in early November 2004, under instruction from Mission Control, the orbiter will perform a propulsive manoeuvre to line itself up and delay its arrival at Titan so that it will be in the right position to view the region where the probe will descend. Data transmitted by the Huygens will be collected by Cassini and stored in its solid-state memory for later transmission to Earth.

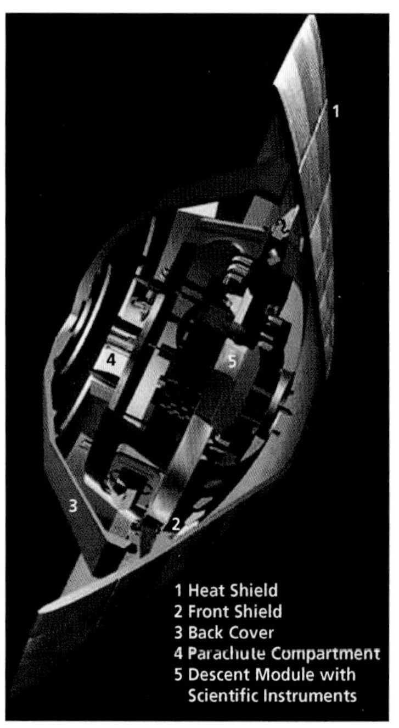

A cutaway drawing of the Huygens spaceprobe, which in 2004 will be the first object to land on the moon of another planet.

As Huygens breaks through the cloud deck, a camera will capture pictures of the Titan panorama. Other instruments will directly measure the organic chemistry in Titan's atmosphere – it will be as if we have been able to despatch a time machine to examine the chemistry of the early Earth. Instruments will also be used to study the properties of Titan's surface during the probe's descent, and perhaps even directly after landing on the surface. But whether the probe will be able to see anything once it gets below the clouds is a moot point – it is equipped with lights, which will help, but even if the skies below the clouds are relatively clear then getting a good long view across the surface will not be easy. At Saturn, the Sun is only a quarter as bright as it is at Jupiter, and light levels on the surface of Titan will be perhaps "equivalent to a gloomy evening" on Earth, says Bob Mitchell. "However, the probe has a rather unusual camera. It has three different detectors that look out from the side at different angles. So it will sweep out a panoramic 360° view all the way down through the atmosphere, and depending on what we find when we actually land, the cameras will hopefully continue to function on the ground. And there's a light on board that will come on just before landing, so we should at least be able to image the surface."

The images beamed back by Huygens will be among the most eagerly awaited from any planetary lander. On Mars and the Moon

This is an artist's view of the centre of Herschel crater, which occupies a large portion of Mimas' leading hemisphere. The near ice formations comprise the central mountains of the crater, with the crater walls visible in the distance.

we at least had a rough idea of what to expect – rocks. On Venus, rocks too, albeit rather hot ones (the fact you are even able to see on Venus at all came as a delightful surprise to the Russian team who built the Venera landers). On Titan, though, who knows? With a surface probably made of frozen and liquid hydrocarbons, water ice and maybe a smattering of silicate rocks, together with lakes or oceans of liquefied methane, Titan will be unlike anywhere visited by humanity's envoys so far.

Bob Mitchell says there are a number of possibilities as to where the probe could end up. "When it lands, we'll probably see a pretty gloomy day down there. It might be more like late twilight [on Earth]. But I'm pretty confident that you could see well enough to walk around and not stumble on anything. As to what we'll see, I guess I really don't have any idea about that. We'll see something that we haven't even thought about, that hasn't even occurred to us, and that's really the most exciting thing about it – that is, after all, the main reason we're going. But whether we find the surface to be all solid ice, or whether we'll find it to be all liquid and we land with a splash, I don't really have any idea."

The possibility that Huygens could be the first probe to splash-land on another planetary body has not escaped the project scientists. Huygens could even end up on a chilly shoreline. ESA's director of science, Roger Bonnet, says, "We might try to aim it at a beach, if we can get some clearer idea of where the continents are, and where the edges of the oceans are – if there are oceans there, of course. That would be fantastic, for we would get pictures of the land and the sea." Landing in, rather than next to, an ocean would be interesting but would quickly curtail the mission. "Huygens is not designed to float," Mitchell says, "however, the probe is equipped with microphones so there remains a small but real possibility that we'll be able to hear the sounds of alien waves lapping against the probe – an intriguing and surreal possibility."

The 1997 launch of Cassini was nearly scuppered – not by technical hitches as with Galileo, but by legal challenges from environmentalists who feared that the spacecraft could potentially release radioactive debris into the atmosphere should the launch fail.

Saturn, Cassini's target, is photographed here by the Hubble Space Telescope. The bright flashes near the saturnian south pole are electtrical discharges above the atmosphere – Saturn's southern lights.

Huygens will be just one highlight of the Cassini mission; the cameras on the orbiter are due to send back over 300,000 colour images of the Saturn and its moons. Like Galileo, the Cassini orbiter will undertake a comprehensive exploration of planet, rings, moons and magnetosphere. During the four-year primary mission Cassini will undertake a tour of the Saturnian system, tracing out sixty orbits of various orientations. Cassini will rely on the gravitational well of Titan to give it a velocity boost to change orbit. If Cassini is still in good health at the end of the primary mission, an extended mission will begin, just as with Galileo. Possibilities for the extended mission include putting the spacecraft in orbit around Titan – the first time a probe would be in orbit around the moon of another planet. This could be achieved by flying Cassini close in to Titan, skimming its atmosphere and using air friction to slow it down sufficiently for it to be captured by the moon's gravity – a technique known as aerobraking, and first used successfully by the Magellan Venus orbiter in 1993.

Problems looming: Shades of Galileo?

In October 2000 Cassini returned its first images, in black and white and in colour. They were of Jupiter, taken from a distance of 84 million km as it closed in upon the giant planet in time for its new year rendezvous with Galileo. JPL announced, quite justifiably, that the pictures represented a great success. "This has been our first opportunity to exercise the Cassini flight and ground systems in a mode very similar to how we expect to operate at Saturn, and I'm extremely pleased with how it is working", said Bob Mitchell in a press statement. "The spacecraft is steadier than any spacecraft I've ever seen", added Carolyn Porco of the University of Arizona, team leader for the camera on Cassini. "It's so steady: the images are unexpectedly sharp and clear, even in the longest exposures taken and in the most challenging spectral regions." Ironically, the first Cassini picture was taken in early October, two months before its December arrival, mirroring Galileo's earlier approach to its target planet. The Cassini picture must show something similar to the one that is still at the beginning of Galileo's tape, forever lost after the tape recorder jammed.

Beautiful pictures of Jupiter aside, it soon became clear that all was not well with Cassini. At the same time as the Jupiter images were released, JPL admitted that a problem had cropped up with the communications link between the orbiter and the Huygens probe. The hitch, which was identified in early September 2000, emerged after tests conducted at the European Space Agency's Operations Centre at Darmstadt, Germany, on the radio receiver supplied by ESA to receive signals from the probe as it descends through Titan's atmosphere. According to the test results, the signal sent to Cassini from Huygens will change in frequency as both spacecraft rapidly change position in relation to each other, much as the Doppler effect seems to change the pitch of a whistle on a moving train. The receiver on Cassini orbiter would not be able to collect all the data from Huygens.

According to NASA, "the best minds in the business are working on solutions". This sounds ominously like the problem all those years ago with Galileo's High Gain Antenna but Bob Mitchell insists there is no cause for alarm. "I don't think we're going to lose any pictures from Huygens", he said after the announcement. "If we did nothing, we

Pluto remains the only planet never to have been visited by a spaceprobe. NASA has plans to send a mission to this remote world which should arrive sometime between 2010 and 2020, but in 2000 the mission was abruptly cancelled. Now, however, it looks as though a mission to Pluto may be back on the cards.

would lose a significant amount of probe data. But we are not going to do nothing. I am optimistic."

The only tentative mission on NASA's books to explore the Solar System beyond Saturn is the Pluto-Kuiper Express. Pluto is the one planet never visited by a space probe, and we know almost nothing about it, or its large moon, Charon. In 1999 there were even moves to reclassify Pluto as a Kuiper Belt Object, changing its status to the largest of the icy worldlets swarming around the outer limits of the Solar System. This demotion – which caused much amusement in the press and indignation among astrologers, who had had to redraw their charts since the discovery of Pluto in 1930 – was rejected by the International Astronomical Union, so Pluto stays a planet. Pluto is probably made largely of frozen gases and water ice, and is deep-frozen at around minus 225°C. There must be very little going on there, no oceans and no active geology at all – though the planet does develop a temporary nitrogen atmosphere for the 60 or so years of its eccentric 240-year orbit during which it is closest to the Sun. Pluto may be one of the best-preserved relics of the earliest days of the Solar System, and for that reason many scientists would very much like to go there. But it looks as though they will have to wait. On 13 September 2000 a NASA "stop work order" was issued for the Pluto-Kuiper Express mission "as currently envisioned". A statement read, "Further direction from NASA has been given to develop a new mission to reach Pluto before 2020". For now, Cassini and Galileo may represent the high-water mark of humanity's exploration of the Solar System.

The Cassini spacecraft took this extraordinary picture in December 2000. It shows Io apparently floating above Jupiter's swirling clouds.

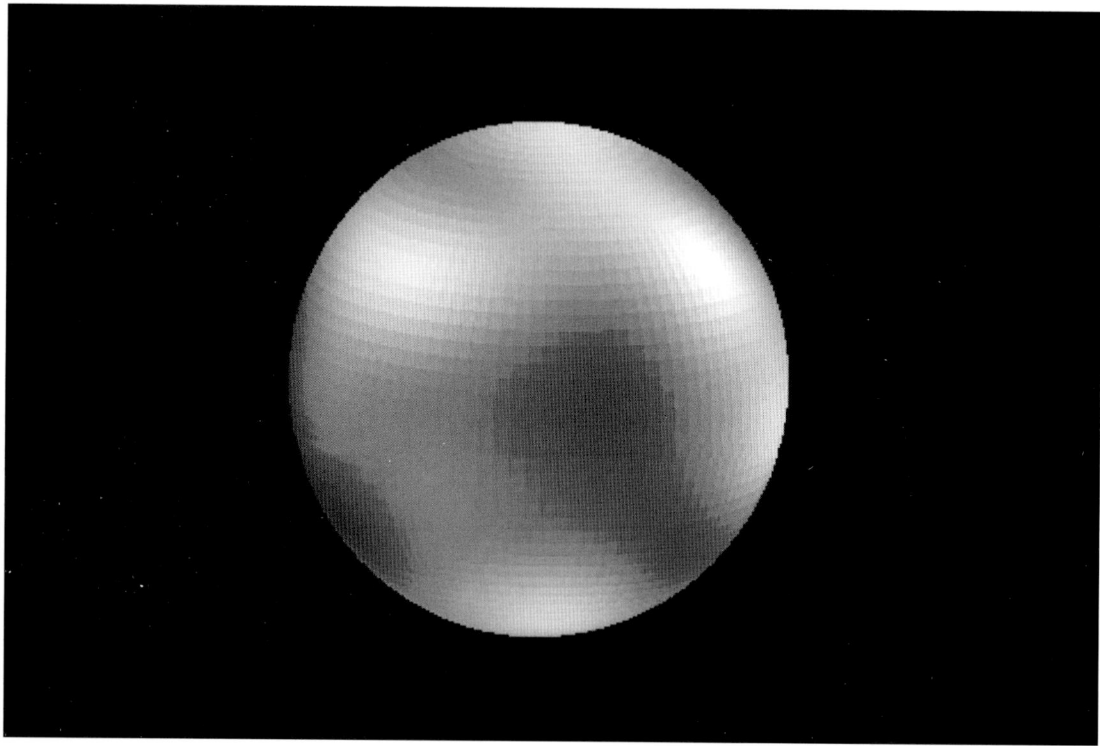

Pluto is the only planet yet to be visited by a spaceprobe. Until JPL gets the green light for a mission to this frozen world and its moon Charon, this fuzzy and indistinct image of Pluto's surface taken by Hubble is the best we have.

GLOSSARY

ammonia A chemical compound present in the atmosphere of Jupiter and other *gas giants*, formula NH_3.

anticyclone A region of high pressure in a planet's atmosphere. An anticyclone rotates clockwise in a planet's northern hemisphere, and anticlockwise in the southern.

asteroid A small and rocky celestial object in orbit around the Sun. Most are concentrated between the orbits of Mars and Jupiter in a zone known as the asteroid belt.

atmosphere The shell of gas around a planet or moon.

aurora A glow in a planet's upper atmosphere caused by the interaction of charged particles from the *solar wind* captured by the planet's magnetosphere and colliding with atoms in the atmosphere.

bar A unit used to measure atmospheric pressure, equal roughly to the average pressure at the Earth's surface.

basin A very large impact *crater*.

billion In this book, a thousand million, or 10 to the power of nine (10^9).

caldera A large volcanic crater caused by the collapse of an underground magma chamber.

charge-coupled device (CCD) An array of light-sensitive elements that produce electricity when hit by photons. A CCD forms the "retina" of Galileo's camera.

comet A small, icy object in orbit around the Sun. Unlike *asteroids*, comets are composed of volatile ices and dust, and when close to the Sun they warm up and produce visible streams of gas known as tails.

compression The compacting of data so that it takes up less space in a computer's memory and is faster to transmit.

core The innermost part of a planet or moon.

crater A generally bowl-shaped depression on the surface of a planet or *asteroid*. Most craters are formed by the impact of an asteroid or meteorite, but some are the result of volcanic action.

crust The solid outer layer of a planet or moon.

differentiation The separation of the interior of a planet or moon into concentric shells. Differentiation happens when a body's interior melts during the early stages of its formation, allowing denser materials to sink and form a *core*, and less dense materials to rise and form a *mantle* and *crust*.

Doppler effect The apparent shift in the frequency of sound waves or electromagnetic waves as the source of the waves moves towards or away from the observer. The Doppler effect was utilized by Galileo to probe the interiors of Jupiter's moons.

erosion The gradual wearing away of a planetary body's landscape, for example by the action of wind, rain and rivers.

flyby The part of a mission during which a space probe passes close by a planet or moon.

Galileo's orbital tour features several flybys of Jupiter's moons.

Galilean satellites The four large moons of Jupiter, discovered by Galileo in 1610. They are, in order of increasing distance from the planet, Io, Europa, Ganymede and Callisto.

gas giant A planet that, in contrast to the *terrestrial planets*, is large made mostly of gases such as *hydrogen* and *helium*; also known as a giant planet. In our Solar System, the gas giants are Jupiter, Saturn, Uranus and Neptune.

geyser A water vent that erupts periodically. The liquid is boiled by hot rocks underground.

geological time "deep" time, the calendar of the Earth and *Solar System*'s formation and evolution, measured in millions and billions of years.

Great Red Spot huge weather system on Jupiter, which has been visible for 300 years

greenhouse effect The warming of an atmosphere brought about by certain gases absorbing solar radiation that is absorbed and re-emitted by the planet, thus prevented it from escaping. On Earth, the principal greenhouse gas is water vapour.

helium After *hydrogen*, the commonest gas in the Universe, and a constituent of stars and *gas giants*.

High Gain Antenna (HGA) Galileo's main communications dish. In April 1991 it failed to open, forcing project engineers to rely on the *Low Gain Antenna*.

hydrogen A gaseous element that makes up the bulk of the Universe and provides the raw material for stars. It is a major constituent of *gas giants*.

ice The solid form of volatile materials – substances, such as water and ammonia, that are liquids or gases at the temperatures prevailing in the inner Solar System. Water ice, which at the temperatures in the vicinity of Jupiter is as hard as granite, is a major component of Ganymede, Callisto and Europa.

ion An electrically charged atom or molecule.

Jet Propulsion Laboratory (JPL) The establishment at Pasadena, near Los Angeles, responsible for developing and controlling unmanned space missions. It is run jointly by NASA and the California Institute of Technology (Caltech).

Jovian Of or pertaining to the planet Jupiter.

lava Molten rock on the surface of a planet.

Low Gain Antenna (LGA) The secondary communications dish on Galileo that was used when the *High Gain Antenna* (HGA) failed.

magma Molten rock underground.

magnetometer An instrument used to measure magnetic fields.

magnetosphere The volume of space around a planet or moon in which particles are governed by the object's magnetic field rather than that of the Sun or another body.

mantle The part of a planet or moon between the *core* and the *crust*.

meteorite A stony or metallic object that strikes a planet or moon.

methane A chemical compound present in the atmosphere of Jupiter and other *gas giants*, formula CH_4.

moon Another name for a natural *satellite*.

National Aeronautics and Space Administration (NASA) A Federal agency concerned with aeronautics and space exploration, formed to counter the perceived Soviet superiority in space.

near-infrared The part of the electromagnetic spectrum just beyond the red end of the visible region.

newton The standard unit of force, defined as the force that gives a 1 kg mass an acceleration of 1 metre per second per second.

noble gas One of six very unreactive chemical elements: helium, neon, argon, krypton, xenon and radon. They were formerly called rare gases or inert gases.

olivine A dark green silicate mineral, rich in metals, found in alkaline igneous rock (rock of volcanic origin).

orbit The path followed by a body moving in a gravitational field which causes a centrally directed force.

orbital tour Name given to the series of elliptical orbits of Jupiter undertaken by the Galileo *orbiter*. Most of these separate orbits took the spacecraft close to one of the *Galilean satellites*.

orbiter The part of a spacecraft put into orbit around a planet or moon, as opposed to a module that is sent to the surface.

organic compound A chemical compound that contains carbon (excluding oxides and carbonates). Organic compounds are so named as they feature in the chemistry of life.

particles and fields experiments Experiments that provide data on magnetic fields, radiation and ionized particles in space.

photodissociation The breakdown of chemical compounds by light energy.

photometer An instrument used to measure luminous intensity.

pixel The smallest unit of a digital picture, a contraction of "picture element".

planet One of the nine large roughly spherical objects that orbit the Sun. Strictly, these are known as major planets, as opposed to minor planets – another term for asteroids.

planetesimal A small object that formed from material that condensed out of the *solar nebula*. Planetesimals provided the building blocks for planets, moons and *asteroids*.

plasma A high-temperature state of matter, resembling a gas, that consists of electrically charged particles – *ions* and electrons.

plate tectonics The "unifying theory" of terrestrial geology, which states that the crust is consists of a series of plates that move relative to one another, giving rise to volcanoes, earthquakes and mountains along their boundaries.

plume geyser-like jet of material seen erupting from the surface of Io.

pyroxene A silicate mineral containing calcium, iron and magnesium.

satellite Any small object in orbit around a larger object, commonly called a moon. The Moon is the satellite of the Earth; the planets are satellites of the Sun. Satellites may be artificial – Galileo is now a satellite of Jupiter.

speed of light In a vacuum, light and other electromagnetic radiation travels at precisely 299,792,456 metres per second.

solar wind The stream of charged particles constantly being ejected from the Sun's surface.

solar nebula The cloud of gas and dust out of which the Sun, the planets and the rest of the Solar System formed, some 4.5 billion years ago.

Solar System The Sun together with the planets and their moons, asteroids, comets and other pieces of assorted space debris.

solid state Describing a data storage or electronic processing system that has no moving parts. A RAM memory chip is a solid state device, whereas a tape recorder is not. Galileo's camera is a solid state imaging system.

spectrometer An instrument that determines the composition of an object from a distance by measuring the wavelengths of light or other electromagnetic radiation reflected off it or emitted by it.

sulphur A chemical element which is a soft yellow crystalline solid at room temperature on Earth. Sulphur and sulphur compounds are found in large quantities on the surface of Io.

terminator The line dividing the day and night sides of a planet or moon.

terrain A particular type of planetary landscape.

terrestrial planet A planet which, in contrast to the *gas giants*, is small and made mostly of rock. In our Solar System Mercury, Venus, Earth and Mars are the terrestrial planets; Pluto was formerly classified as one.

trillion A million million, or thousand *billion*.

Titan rocket A conventional expendable launch vehicle built for **NASA** by Lockheed.

ultramafic An igneous rock with a mineral composition typical of the deep crust or mantle, consisting almost exclusively of iron and magnesium silicates and little quartz or feldspar.

volcanism The eruption of molten rock and/or gases onto the surface of a planet.

watt The standard unit of power, defined as the expenditure of 1 joule of energy per second.

BIBLIOGRAPHY

Books

BELTON, MICHAEL (ed.) *Cruise to Jupiter: Contributions of the Galileo Imaging Science Team.* Galileo Project Document 625-800, JPL, 1996.

BELTON, MICHAEL (ed.) *In Orbit at Jupiter: Contributions of the Galileo Imaging Science Team.* Galileo Project Document 625-801, JPL, 1996.

Galileo Project Team *Galileo: The Tour Guide.* Jet Propulsion Laboratory and California Institute of Technology, 1996.

HARLAND, DAVID *Jupiter Odyssey*. Springer Praxis, 2000.

KLUGER, JEFFREY *Journey Beyond Selene*. Little, Brown, 1999.

PANEK, RICHARD *Seeing and Believing*. Fourth Estate, 2000.

SOBEL, DAVA *Longitude*. Fourth Estate, 1996.

Articles

Earth and Venus

BELTON, MICHAEL *et al.* 'Images from Galileo of the Venus cloud deck'. *Science*, vol. 253, pp. 1531–6, 1991.

SAGAN, CARL *et al.* 'A search for life on Earth from the Galileo spacecraft'. *Nature*, vol. 365, pp. 715–21, 1993.

TOIGO, ANTHONY *et al.* 'High-resolution cloud feature tracking on Venus by Galileo', *Icarus*, vol. 109, pp. 318–36, 1994.

Asteroids

BELTON, MICHAEL *et al.* 'First images of asteroid 243 Ida'. *Science*, vol. 265, pp. 1543–7, 1994.

BELTON, MICHAEL *et al.* 'The discovery and orbit of 1993 (243)1 Dactyl'. *Icarus*, vol. 120, pp. 185–99, 1996.

GREENBERG, RICHARD *et al.* 'Collisional history of Gaspra'. *Icarus*, vol. 107, pp. 84–97, 1994.

VEVERKA, JOSEPH *et al.* 'Galileo's encounter with 951 Gaspra: Overview'. *Icarus*, vol. 107, pp. 2–17, 1994.

Jupiter

CHAPMAN, CLARK et al. 'Preliminary results of Galileo direct imaging of S-L 9 impacts'. Geophysical Research Letters, vol. 22, pp. 1561–4, 1995.

GIERASCH, PETER *et al.* 'Observation of moist convection in Jupiter's atmosphere'. *Nature*, vol. 403, pp. 628–30, 2000.

SEIFF, ALVIN 'Dynamics of Jupiter's Atmosphere'. *Nature*, vol. 403, pp. 603–5, 2000.

Satellites (general)

GRIFFITH, CAITLIN A. 'Detection of daily clouds on Titan. *Science*, vol. 290, pp. 509–13, 2000.

MCKINNON, WILLIAM 'Galileo at Jupiter – meetings with remarkable moons'. *Nature*, vol. 390, pp. 23–6, 1997.

NEUBAUER, FRITZ 'Oceans inside Jupiter's moons'. *Nature*, vol. 395, pp. 749–51, 1998.

Thomas, P.C. *et al.* 'The small inner satellites of Jupiter'. *Icarus*, vol. 135, pp. 360–71, 1998.

Europa

CARR, MICHAEL *et al.* 'Evidence for a subsurface ocean on Europa', *Nature*, vol. 391, pp. 363–5, 1998.

CHAPMAN, CLARK 'What's under Europa's icy crust?' *Technology Today*, Autumn 1998.

GEISSLER, PAUL *et al.* 'Evidence for non-synchronous rotation of Europa'. *Nature*, vol. 391, pp. 368–70, 1998.

GREELEY, RON 'The partially watery world of Europa, one of Jupiter's moons'. *Earth in Space*, vol. 10, no. 4, pp. 1–16, 1997.

GREELEY, RON *et al.* 'Europa: Initial Galileo geological observations'. *Icarus*, vol. 135, pp. 4–24, 1998.

GREENBERG, RICHARD 'Europa's lifelines', *New Scientist*, November 2000.

KIVELSON, MARGARET *et al.* 'Galileo magnetometer measurements: A stronger case for a subsurface ocean at Europa'. *Science*, vol. 289, pp. 1340–43, 2000

PAPPALARDO, BOB *et al.* 'The hidden ocean of Europa', *Scientific American*, October 1999.

Io

BELTON, MICHAEL *et al.* 'High-temperature hotspots on Io as seen by the Galileo solid state imaging (SSI) experiment'. *Geophysical Research Letters*, vol. 24, pp. 2443–6, 1997.

GRAPS, AMARA *et al.* 'Io as a source of the Jovian dust streams'. *Nature*, vol. 395, 2000.

JOHNSON, TORRENCE 'Io'. In *Contributions of the Galileo Imaging Team*, JPL internal publication, 1999.

LOPES-GAUTIER, ROSALY 'Volcanism on Io'. In *Encyclopaedia of Volcanoes*, Academic Press, pp. 709–23, 2000.

MCEWEN, ALFRED *et al.* 'Active volcanism on Io as seen by Galileo SSI'. *Icarus*, vol. 135, pp. 181–219, 1998.

McEWEN, ALFRED *et al.* 'High-temperature silicate volcanism on Jupiter's moon Io'. *Science*, vol. 281, 1998.

McEWEN, ALFRED *et al.* 'Galileo at Io: Results from high-resolution imaging'. *Science*, vol. 288, 2000.

WILSON, LIONEL 'Io writes its history in hot metal'. *Nature*, vol. 394, pp. 520–21, 1998.

Ganymede and Callisto

KIVELSON, MARGARET *et al.* 'Discovery of Ganymede's magnetic field by the Galileo spacecraft'. *Nature*, vol. 384, pp. 537–41, 1996.

KIVELSON, MARGARET *et al.* 'Induced magnetic fields as evidence for subsurface oceans in Europa and Callisto'. *Nature*, vol. 395, pp. 777–80, 1998.

PAPPALARDO, ROBERT 'Ganymede and Callisto'. In *Contributions of the Galileo Imaging Science Team*, JPL internal document, 1998.

PAPPALARDO, ROBERT *et al.* 'Grooved terrain on Ganymede: First results from Galileo high-resolution imaging'. *Icarus*, vol. 135, pp. 276–302, 1998.

Internet sources

The Galileo Project Homepage: www.jpl.nasa.gov/galileo/

The JPL Homepage: www.jpl.nasa.gov

Latest news from NASA: www.nasa.gov/today/index.html

The Cassini Project Homepage: www.jpl.nasa.gov/cassini/

For details of other missions, and JPL's plans for future Solar System exploration, see: http://spacescience.nasa.gov/missions/index.htm

INDEX